D1034891

Criminal Justice
Recent Scholarship

Edited by
Marilyn McShane and Frank P. Williams III

A Series from LFB Scholarly

Environmental Crime and the Media
News Coverage of Petroleum Refining Industry Violations

Melissa L. Jarrell

LFB Scholarly Publishing LLC
New York 2007

Library of Congress Cataloging-in-Publication Data

Jarrell, Melissa L. (Melissa Luisa), 1976-
  Environmental crime and the media : news coverage of petroleum
refining industry violations / Melissa L. Jarrell.
     p. cm. -- (Criminal justice recent scholarship)
  Includes bibliographical references and index.
  ISBN-13: 978-1-59332-205-2 (alk. paper)
  ISBN-10: 1-59332-205-4 (alk. paper)
  1. Petroleum industry and trade--Press coverage--United States. 2.
Offenses against the environment--Press coverage--United States. I.
Title.
  PN4888.P47J37 2007
  070.4'4936414--dc22

                                              2006102798

ISBN-10 1593322054
ISBN-13 9781593322052

Printed on acid-free 250-year-life paper.

Manufactured in the United States of America.

Table of Contents

Acknowledgements

In loving memory of my grandfather, Millard Jarrell and to my wonderful Nana, Betty Jarrell, for providing financial and emotional support throughout my cherished life. Dad, thank you for everything; I love you more than words can say. Mom, you have always been an inspiration to me. You have unselfishly given your time and energy to the dying, the sick, and the homeless. I champion the downtrodden because you taught me to help those less fortunate than us. To Mike Lynch, my mentor and major professor. Your achievements are remarkable. I consider myself lucky to have learned from you. To Danielle, my long-distance partner in crime and to Katie, my dearest friend and soul mate. To Matilde Marines, for her unending patience and support. Thank you!

CHAPTER 1

Environmental Crime

INTRODUCTION

Environmental crime is considered to be one type of white-collar or corporate crime (Burns and Lynch, 2004). The government, the media, and the public rarely conceptualize environmental harm and injustice as "crime" despite overwhelming evidence to the contrary. Chapter One provides an overview of environmentalism in the United States; describes the nature and impact of environmental crime; explores the causes and consequences of environmental crime and injustice in terms of environmental and human impacts; presents federal environmental legislation and enforcement efforts with a discussion of the various problems associated with legislation and enforcement, and highlights corporate and political responses to environmental crime and legislation. This chapter also presents an overview of studies of environmental crime within the criminological literature.

Environmentalism in the United States

A majority of the American public, if asked about the roots of environmentalism, would likely recall the 1970s and the back-to-basics mentality of the "hippies". They may even remember the first celebration of Earth Day. While the large-scale environmental movement may indeed be rooted in the 1960s and 1970s, environmental awareness and resource conservation have been hot topics throughout American history. In order to understand current environmental legislation, debates, and controversies, it is critical to understand what has led us to where we our today.

Although heightened concerns for environmental crime and justice are relatively recent, public and political concern for the environment is not a new phenomenon. At times throughout our nation's history, environmental issues have received a great deal of attention publicly and politically, while at other points in history, our nation has focused its attention on arguably more pressing issues including national security, economic development, and the like.

Waves of Concern

According to Taylor (2000), environmentalism in the nineteenth century was characterized by the "exploitative capitalist paradigm". Environmental destruction was considered to be "the inevitable by-product of growth, consumption, and industrial advancement" (Taylor, 2000: 529). However, toward the turn of the twentieth century, environmentalists cautioned that with rapid urbanization and industrialization, our nation must focus on preserving nature and protecting species for the enjoyment of later generations. For the most part, environmental concern focused on conservation and protection of natural spaces. Leaders of the "romantic environmental paradigm", including Rosseau, Muir, Marsh, and others, challenged people to protect the wilderness and live harmoniously with nature (Taylor, 2000).

During the 1960s, however, environmental problems affecting human health were becoming increasingly apparent. Rachel Carson's book *Silent Spring* (1962) highlighted industrial and government practices with respect to the use of pesticides and other chemicals. Citizen groups grew in number and fought for more governmental involvement in the regulation of environmental hazards. The "new environmental movement" evolved from the social fervor of the 1960s (Taylor, 2000) and involved legal-scientific groups and the creation of the Natural Resources Defense Fund (NRDC), the Sierra Club Legal Defense Fund (SCLDF), and the Environmental Defense Fund (EDF) (Cole, 1992). In early 1970, Congress authorized the creation of a federal agency to oversee federal anti-pollution efforts and administer regulatory and environmental protection legislation; the Environmental Protection Agency (EPA).

Although the "new environmental movement" (Taylor, 2000) can be credited with aiding in the creation and implementation of federal environmental legislation aimed at preserving nature and regulating pollution, their efforts were not intended to achieve equitable justice. The "new environmental movement", comprised mainly of college-educated, white, middle class activists and legal scholars with a growing enthusiasm for outdoor recreation, was more concerned with resource conservation than with human environmental hazards and more inclined to strive for small systematic changes rather than radical political and social changes. The "environmental justice paradigm" (Taylor, 2000) involves activists who are most directly affected and more severely affected by environmental problems (Cole, 1992). Mainstream or "new" environmentalists are primarily concerned with aesthetic and recreational considerations; are overwhelmingly white and middle class, use litigation for problem solving; and typically pinpoint a single bad actor as the cause of an environmental problem. In contrast, grassroots activists are often fighting for health and home; are primarily poor and working class people of color; often have a greater distrust of the law and more experience with non-legal strategies; they adopt a social justice orientation that calls for structural level changes so as to address the deeper problems of poverty, crime, unemployment, and environmental destruction.

> "People living in or near industrial communities know that law-abiding and law-breaking corporations differ in degree only; both put pollutants out the smokestack, and both thus poison nearby communities. In contrast to the bad actor model, which seeks to identify and punish individual bad actors, the institutional model identifies individual polluters not as explanations themselves' but merely as part of an overall system of maximizing profit" (Cole, 1992: 25).

While mainstream environmentalists view pollution as the failure of government and industry to clean up the mess of a few violators, grassroots activists see pollution as the *success* of industry in maximizing profits by externalizing environmental costs (Cole, 1992). Similar to deeming unemployment as the failure of the individual rather than our economic system, many regard environmental crime as the result of a few corrupt individuals rather than the result of our capitalist economy (Mayo and Hollander, 1991).

What is Environmental Crime and Injustice?

Most definitions of environmental crime consider such crime to cover acts or omissions that violate federal, state, or local environmental standards and laws (National White Collar Crime Center, 2004; Situ and Emmons, 2000). Some acts, especially those committed by corporations, may not violate the criminal law. Many are violations of regulatory laws (Burns and Lynch, 2004). Many of these acts cause a great deal of harm to the environment and human health and safety and therefore should be treated as criminal (Clinard and Yeager, 1980; Frank and Lynch, 1992; Lynch, 1990; Reiman, 1998).

Environmental crime typically affects many victims and the victimization may be gradual and/or silent (Frank and Lynch, 1992). The Federal Bureau of Investigation (2003) focuses its attention on the most serious threats to public health and natural resources such as cases involving handling of hazardous waste and pollutants that may endanger workers, environmental catastrophes that place entire communities at risk, federal government facility violations, businesses identified by regulatory agencies as having a long history of violations or flagrant disregard for environmental laws and organized crime activities generally in the solid waste industry.

A number of studies in disciplines outside of criminology have examined victimization distributions by examining exposure to toxic hazards in relation to community characteristics. For example, it has been shown that minority or low-income communities are disproportionately affected by environmental hazards (Bullard, 1983; Lavelle and Coyle, 1992; Mohai and Bryant, 1992; United Church of Christ, 1987; U.S. General Accounting Office, 1983). The majority of these studies fall within an area of research identified as "environmental justice." Environmental justice advocates argue that no person, regardless of race, class, or gender, should suffer the consequences of environmental degradation and therefore substantial political, social, and economic efforts should be made to protect the environment and human health.

Conceptualization of Environmental Crime and Injustice

Environmental crime often goes unnoticed and people are somewhat apathetic to the problems caused by environmental crime. For the most part, the apathetic response to environmental crime is a direct result of public unawareness of the real dangers to health and safety posed by this type of criminal behavior. Most environmental hazards commanding political, public, and media attention have been those hazards that can be easily "pinpointed" at particular places and locations and where cause and effect could be closely linked (Mayo and Hollander, 1991). "Some hazards remain hidden or unattended because they lie embedded in a societal web of values and assumptions that either denigrates the consequences or deems them acceptable, elevates associated benefits, and idealizes certain notions, or beliefs" (Mayo and Hollander, 1991: 12).

Causes of Environmental Crime and Injustice

Environmental crimes are usually committed for economic reasons and more often than not, corporations place the value of money over public health. Criminal pollution is an economic crime committed to escape costs of dealing with things properly. "If compliance expenses are costly, and the chances of being caught are minimal, a strong incentive to pollute exists. Especially, if they can be reasonably sure that the penalty that will be imposed will be a monetary one in the way of the fine" (Albanese and Pursley 1993: 317). In order to be a deterrent, the penalties must outweigh the crime. Fines are related to the offense, not the offender so often small companies pay too much and super-rich corporations a drop in the bucket (Wilson, 1986).

Environmental crime and environmental injustice then are a result of industry and corporate decisions to maximize profits and externalize costs. In addition, environmental injustice and environmental racism are a result of the political and economic processes that exist at all levels of government. State governments must balance economic development and community interests in health and safety. These decisions are often influenced by corporate donations. On the federal level, members of Congress as well as Presidential candidates are courted by industry and given massive campaign contributions in order to affect legislative issues (Pope, 2004). Often, economic and political decisions at the state and federal level benefit industry or a segment of the community at the expense of others in the community. "Grassroots activists, whose homes are being contaminated or who want to prevent a chemical plant from locating next to them, . . .they ask for help from federal agencies, like the EPA. The EPA, which is under intensive pressure from legislators who are in support of wealthy national and transnational corporations, is caught in the middle of a contentious political fight" (Roberts and Toffolon-Weiss, 2001:76). Environmental crime and injustice then, is the result of corporate and political decision-making that appears to benefit a few at a substantial cost to many.

Impact of Environmental Crime and Injustice

The devastating effects of environmental crime are not easy to determine or estimate. Although we gather and report a wide range of statistics of street and violent crime at the national, state, and local level, there are virtually no uniform or national statistics describing the status and impact of environmental crimes. There is a continuing debate over the consequences and extent of environmental pollution and serious questions over the impact of enforcement on the nation's competitiveness in global/domestic markets. The cost of environmental toxic abatement and clean up is often more than the government is willing to spend. Researchers suggest that environmental crime causes more illness, injury, and death than street crime (Burns and Lynch, 2004; Albanese and Pursley, 1993). According to Burns and Lynch (2004: ix), "we estimate that each year in the United States, up to ten times as many people die from environmental crimes, such as exposure to toxins in the workplace, home, and school, as die by homicide."

Particular problems are presented by the research produced by "corporate" scientists (Lynch and Stretesky, 2001). Despite the abundance of toxic chemicals everywhere, it is difficult to establish a direct causal link between adverse health effects and chemical contamination (Adeola, 2000). "Establishing the time, space, and non-spurious causality of ailments of individuals due to their exposure to toxic chemicals has been the pivotal issue in numerous cases" (Adeola, 2000: 6). Affected communities strongly believe that their symptoms are not taken seriously and are blamed on unhealthy lifestyles. Evidence to support residents' claims is even more difficult to obtain because there is a lack of sufficient baseline data on their health prior to the arrival of industry. Health assessments are extremely expensive and time-consuming and not without methodological flaws. Yet, as Bryant (1995: 9) notes "Although we may not be able to prove causality due to confounding variables such as smoking, diet, indoor pollution, and synergistic and repeated effects of multiple exposures, this does not mean that cause and effect does not exist; it may mean only that we failed to prove it". Bryant summarized the problem as follows: "When the burden of proof is on the community to demonstrate certainty, policy makers often want to hold them to the rigors of traditional research. Yet, when policy makers initiate siting and remediation decisions, they often fail to apply that same level of rigor for certainty as they do for community groups" (Bryant, 1995: 14).

Environmental Impact

Since the beginning of the Industrial Revolution, 200 billion tons of carbon dioxide have been added to the atmosphere (Owen, 1975). Nearly 70,000 chemical products have been introduced since World War II and 1,500 are added each year. The total U.S. production of chemicals amounts to over 300 million tons annually (Goldman, 1991). The Environmental Protection Agency has identified over 700 substances as hazardous to the environment. Over 2 billion pounds of toxic chemicals are released into the environment legally each year (Gray, 1998). According to the EPA, of the 100 billion tons of hazardous waste produced each year in the U.S., 90% is disposed of in an environmentally unsafe manner (Humphries, 1990). The U.S. Department of the Interior and Commerce has identified more than 1,500 species of wildlife as threatened or endangered. In 1995, 46 contaminants, from dioxin to chlordane, were found in fish. The number of lakes, rivers, and other U.S. waterways where consumers have been advised to avoid or limit consumption of trout, salmon, or other species because of chemical contamination rose from 1,278 in 1993 to 1,740 in 1995 (Council on Environmental Quality, 1995).

Human Impact

Each year in the United States, environmental pollutants and hazards are responsible for thousands of illnesses, injuries, and deaths. An estimated 40 million people, one-sixth of the U.S. population, live in close proximity to one or more hazardous waste sites (Cope 2002). Day (1989) argues that polluted water is the single greatest cause of human illness and death through disease. In 1995, over 40 million Americans were served by drinking water systems with lead levels exceeding the regulatory action level (Nadakavukaren, 1995). The EPA states that as many as 30,000 waste sites may pose significant health problems related to water contamination (Department of Health, Education, and Welfare, 1980). More than half the U.S. population lives in counties that violate the Clean Air Act (DeLuca, 1999). Approximately 53,000 people each year die prematurely from lung ailments as the result of air pollution (Situ and Emmons, 2000). Furthermore, Nelkin and Brown (1984) suggest that air pollution kills about 100,000 workers each year and results in 400,000 cases of disease. Cancer death rates are highest in areas close to petrochemical plants, steel mills, and metal refineries (Berry, 1988; Whelan, 1985). According to a 1999 Committee on Health Risks of Exposure to Radon, radon in homes in the United States accounts for 15,400 to 21,800 lung cancer deaths each year; 10% of the total deaths attributed to lung cancer each year (Goldstein and Goldstein, 2002). In 1993, over 40% of the Hispanic population and over 35% of the Asian/Pacific population were exposed to poor air quality (Council on Environmental Quality, 1995). Three out of every five African-Americans live in communities with uncontrolled waste sites (United Church of Christ, 1987).

Residents who believe their health is directly affected by environmental degradation are faced with numerous personal challenges. Not only must residents deal with the health consequences of pollution, they also are subjected to numerous emotional and economic maladies. Their homes lose value, they worry about cancer, and they are concerned about losing jobs in the very industry that pollutes them. Fighting for justice creates even more hardships. When community residents complain about pollution, experts are called in to determine if a problem exists and whether or not is it directly caused by the polluting company. From the perspective of the common citizen, scientific findings are often extremely technical and difficult to comprehend. "Taking the struggle for environmental justice out of the community and into the domain of scientists plays into the domain of risk producers because they have resources and access to scientists" (Kuehn, 1996). For example, in Mossville, Louisiana, residents living near a large chemical plant were found to have abnormally high levels of dioxin in their blood. Experts were called in to ascertain the cause of the abnormalities, while community leaders expressed frustration at being unable to understand the technical reports. One activist sent the following e-mail:

"I, as an average citizen, do not know what half the words. . .from your e-mail means. I do know that many people in Mossville are ill. I personally invite you to come to Mossville, meet the people, and discuss it with those who are affected. Then, perhaps you could go back and find a way to help us instead of playing with words. We are sick. We need help." (Roberts and Toffolon-Weiss, 2001: 19).

Overall, environmental crime and injustice are responsible for a great deal of harm to the environment and human health. Although concern for the environment isn't a new phenomenon, the growth of urbanization and industrialization has increasingly put environmental issues at the forefront of public concern.

Environmental Legislation

The federal government did not enter the field of environmental law until 1890 mainly because political leaders were concerned over their constitutional authority to develop and regulate natural resources (Campbell-Mohn, Breen, and Futrell, 1993). In 1890, Congress established the Rivers and Harbors Act but true federalization of environmental law really took effect about 25 years ago in conjunction with the nation's first Earth Day on April 22, 1970. In July 1970, President Nixon created the Environmental Protection Agency. A great deal of environmental legislation was passed during the 1970s (see Table 1) including: the National Environmental Policy Act (1969), the Clean Air Act (1970), Clean Water Act (1972), Federal Insecticide, Fungicide, and Rodenticide Act (1972), Endangered Species Act (1973), Safe Drinking Water Act (1974), Resource Conservation and Recovery Act (1976), the Toxic Substance Control Act (1976) and the Comprehensive Environmental Response, Compensation, and Liability Act (1980). Today, environmental law encompasses a broad range of federal, state, and local statutes, regulations, and case law relating to the prevention and clean-up of contamination of the environment by chemicals, hazardous waste, and other pollutants.

Table 1: Major Federal Environmental Legislation (1970s)

National Environmental Policy Act (1969)	Enacted to establish a national policy for the environment and provide for the establishment of a Council on Environmental Quality.
Clean Air Act (1970)	Enacted to prevent the deterioration of air quality through controlling emissions of pollutants from sources that cause or contribute to air pollution or endanger human health.
Clean Water Act (1972)	Enacted to restore and maintain the integrity of the Nation's waters and to regulate the sources of water pollution.
Federal Insecticide, Fungicide, and Rodenticide Act (1972)	Enacted for federal control of pesticide distribution, sale, and use in the U.S., for the study of the consequences of use, and to require users to register when purchasing pesticides.
Endangered Species Act (1973)	Enacted to encourage the development and maintenance of conservation programs to safeguard endangered and threatened species.
Safe Drinking Water Act (1974)	Enacted to protect drinking water in the U.S., establish safe standards of purity, and require all owners and operators of public water systems to comply with primary standards.
Resource Conservation and Recovery Act (1976)	Enacted to protect human health and the environment from the dangers associated with waste management and disposal; to encourage the conservation and recovery of natural resources through reuse, recycling, and waste minimization.
Toxic Substances Control Act (1976)	Enacted to regulate chemical substances to which the public or environment may become exposed.
Comprehensive Environmental Response, Compensation, and Liability Act (1980)	Enacted to address problems associated with abandoned hazardous waste sites and clean-up of these sites.

Problems with Environmental Law

Attorneys engaged in environmental law and academics in environmental studies often agree that our environmental legislation is extremely complex and often vague (Lavelle, 1993). For example, the EPA has received so many queries about the meaning of the Resource Conservation and Recovery Act, it set up a special hotline for RCRA questions. In a National Law Journal survey of 200 corporate environmental attorneys, over fifty percent stated that most of their time and energy was spent trying to determine whether or not their companies were complying with the law (Lavelle, 1993). The lack of legislative clarity and complexity has led to a great deal of inconsistent enforcement of such laws.

Enforcement of Environmental Laws

Except for a few highly sensational cases, the criminal prosecution of environmental violations at the federal and state level is a relatively recent development (Edwards et al, 1996). In 1981, the EPA created the Office of Environmental Enforcement and the Department of Justice established an Environmental Crimes Unit. Prior to 1982, only 25 environmental crimes were prosecuted by the federal government (Campbell et al, 1993). Since 1982, the federal government has secured over 1,400 criminal indictments and over 1,000 convictions for violations of environmental law. Since 1974, the courts have assessed over $3 billion in civil and judicial penalties and over $290 million in criminal penalties (Reske, 1992). According to the FBI (2003), at any given time the organization is involved in the prosecution of about 450 environmental crimes cases in conjunction with other federal agencies, in particular the EPA. Over fifty percent of the 450 FBI environmental crimes cases involve violations of the Clean Water Act (FBI, 2003). The FBI generally focuses attention to environmental crimes only when the cases are very serious and involve immediate threats to public health and natural resources (FBI, 2003).

In 1997, U.S. attorneys initiated criminal investigations against 952 individuals or organizations involved in violations of environmental law (BJS, 1999). Approximately one quarter of the suspects were identified as organizations. In 1997, 446 defendants were charged with criminal environmental violation (47% for the unlawful emission of a hazardous substance or other pollutant and 53% for a wildlife violation). Approximately one quarter of those convicted for environmental law violations were sentenced to prison with an average sentence length of 21.5 months (BJS, 1999). Sixty-four percent of those convicted were order to pay a fine and the average fine imposed was $67,416. In that same year, the Federal government filed 207 civil cases involving the violation of environmental laws. Seventy-three percent of these cases ended with a settlement (27%) or a consent agreement (46%) (BJS, 1999). Although federal and state enforcement efforts targeting environmental crime have increased in recent years, only a small percentage of criminal enforcement efforts are aimed at environmental crime.

Problems with Environmental Enforcement

According to Albanese and Pursley (1993), one of the main problems with environmental regulation and enforcement is the fragmented nature of authority. No singular agency is responsible for regulation and enforcement of our federal environmental laws. The Environmental Protection Agency, Nuclear Regulatory Commission, Occupational Safety and Health Association, Department of Energy and the Agency for Toxic Substances and Disease Registry are just some of the many agencies that deal with regulation and enforcement of laws concerning the environment. On the state level, there are several different approaches to dealing with environmental law violations which are all a great deal more complex than dealing with conventional crime. Although state and local prosecutors are given the authority to enforce these laws, they have rarely focused on such violations, due in large part to the overwhelming and complex nature of environmental crime and the public and media focus on conventional crime.

Despite the creation and implementation of new and amended laws to address environmental hazards, it is proving very difficult to enforce such laws (Albanese and Pursley, 1993). Penalties for law violations are generally handled by administrative agencies who impose fines; most actions do not result in criminal penalties. "And unlike most conventional crimes, it is generally impossible to determine the seriousness of the offense by the nature of the action taken" (Albanese and Pursley, 1993: 306). A study of violations of regulatory laws by Fortune 500 companies revealed that even serious violations generally received only administrative sanctions (Clinard and Yeager, 1980).

Corporate and Political Responses to Environmental Crime Legislation

Industry has learned to deal effectively with environmental crime and justice legislation, mandates, and communities. During the 1970s and 1980s, as more and more laws established industrial rules and regulations, industry was faced with a vast bureaucracy and expensive clean-up costs. Industry has not given in to these laws or grassroots campaigns. Rather than create non-polluting alternatives, corporations prepared for war, which made future struggles even more difficult. To challenge environmental justice legislation and mandates, corporations have utilized a wide range of techniques including: the "greenwash" and "spin" of environmental justice claims (Stauber and Rampton, 1995) and "environmental blackmail" (meeting suggested standards will force the industry to move, costing the communities jobs). Industry has spent millions of dollars on PR campaigns to protect their image and promote their new "green" attitudes. Each year, Earth Day is sponsored by the worst polluters in the business. Grassroots activists have been called "insane", "half-cocked nut cases", "extremists" and "opportunists" (Roberts and Toffolon-Weiss, 2001).

Corporations are also fighting for "voluntary standards" and spending billions of dollars on no-holds-barred lobbying and on PR campaigns that present their new, greener image (Lynch and Stretesky, 2001). The EPA has been directly attacked by industry lawyers who argue that the EPA lacks authority under the law to force states to undertake new environmental policies.

The current Bush Administration has effectively reduced many of the victories accomplished during the Clinton Administration. Industry has been successful at the federal level at combating environmental justice. Congressional Republicans placed into the VA-HUD Appropriations Bill in 1998, a one-year moratorium on the EPA's using of any funds "to implement or administer the interim guidelines" for Title IV complaints filed after October 21, 1998 and this moratorium was extended through 1999 (EPA, 2000). According to William Kovacs, vice president for environmental policy for the U.S. Chamber of Commerce, "these appropriations provisions are central to our efforts to stop the worst excesses of the environmental movement. . .this will block the worst kind of environmental lunacy masquerading as civil rights" (U.S. Chamber of Commerce 1998). Environmental justice advocates emphasize that the mandates don't address nearly *enough*; broader problems need to be addressed such as land values, lowered quality of life, and the like (Cushman, 1998).

To address this issue, the EPA put together a 25-member committee to revise the Interim Guidelines with representatives from industry, state/local governments, and the scientific community. The revised guidelines were supposed to be published in 1999 but no consensus was reached. The committee decided to issue a report expressing the various diverging opinions in conjunction with ideas for future decision-making (Sissell, 1999). On June 16, 2000, the EPA issued a Revised Guidance for Investigating Title VI Administrative Complaints Challenging Permits. The revised report made state environmental agencies responsible for determining what constituted a significant disparate impact. Accordingly, only the very worst cases would be subject to compliance.

Environmental Crime and Criminology

In recent years, academic attention to problems of environmental crime and injustice and racism has increased. Scholars in environmental studies, law, public health, and other disciplines have focused their efforts on understanding the problem and offering solutions. Despite this unprecedented growth in other fields of inquiry, criminologists, aside from a few individuals, have been relatively silent.

A handful of criminologists have, in recent years, turned their attention to crime as it relates to the environment. However, "environmental crime anthologies (Clifford, 1998; Edwards et al., 1996) have largely overlooked the social power context in which environmental deviance occurs" (Simon, 2000). "The absence of environmental justice studies in the criminological literature speaks to the unwillingness to take issues of racial and class discrimination and corporate harm seriously" (Lynch, Stretesky, and McGurrin, 2001). Although there has been an unprecedented increase in the number of criminal statutes for environmental crimes, there has been little academic research concerning this form of sanctioning (Edwards et al., 1996). "Criminologists not only neglect the harms caused by corporate crime, but have also neglected the laws which criminalize these behaviors" (Lynch, Stretesky, and McGurrin, 2001).

The few criminologists who have endeavored to study environmental crime have made important contributions to the literature. Stretesky and Lynch (1999) addressed the connections between institutionalized racism and corporate violence. "The problem is learning to accept that when companies dump chemicals into rivers, streams, and landfills, or alongside roadways, they do so purposefully and with knowledge that the likely results of their actions will include injury and death for those exposed to their waste products. These are not accidents- they are planned actions no less serious than assaults of killings" (Stretesky and Lynch, 1999: 169). The authors conclude that corporate environmental violence cannot be alleviated via traditional criminal justice responses. In order to understand and effectively deal with corporate environmental violence, we must acknowledge social structural factors, economic variables, and institutionalized racism as primary causes of this type of violence. Lynch and Stretesky (2001) demonstrate ways criminologists can employ medical evidence to identify toxic harms. In their study, the authors find a great deal of literature supporting the premise that toxic chemicals are directly related to illness and death in the United States. Furthermore, they note the lack of attention given to the production and use of safe alternatives. Research also shows that corporations deny or ignore their role in causing death and illness (Lynch and Stretesky, 2001). The authors conclude that although the link between illness and disease and toxic exposure is well documented in the medical literature, industry continues to manipulate data, spend millions of dollars on elaborate public relations campaigns, and hide important findings.

Conclusion

Environmental crime, in its various forms, is responsible for a great deal of environmental and human harm. Complexities and problems associated with environmental legislation make it difficult to enforce environmental laws. Rather than consider safe alternatives to environmental pollution, industry has spent money on PR campaigns to present to the public an image of "environmental friendliness". For the most part, the public is unaware of the dangers to the environment and human health associated with environmental crime. This public ignorance combined with corporate strategies touting corporate environmental responsibility creates a false and misleading image of environmental crime. The importance of understanding environmental crime across many categories cannot be over-emphasized. Awareness and understanding are the first steps leading to meaningful change. Although a small number of criminologists have begun to examine issues related to environmental crime, in order for this information to reach the public, information must be available outside of the academic literature. One of the primary vehicles for presenting mass information about important social issues is the mass media.

CHAPTER 2

The Mainstream Mass Media

INTRODUCTION

Each day, we are bombarded by a wide range of media messages. It is nearly impossible to avoid the media. Whether or not we choose to watch TV, surf the web, listen to the radio, or read a newspaper, chances are high that we are exposed to the mainstream mass media in some form on a daily basis. Chapter Two presents a historical overview of mass communication in the United States; explores the major differences between the mainstream mass media and alternative media sources, discusses the factors influencing media information; and highlights the political and social impact of mainstream mass media exposure.

What is Mass Communication?

Mass communication is a method by which mediated information is disseminated to a large audience of people. Mass communication differs from interpersonal communication in a number of ways, most notably for its potential for far greater impact than interpersonal communication (Rodman, 2001). Mass communication is synonymous with the mass media. The mass media disseminate information in various forms through a vast number of sources including television, newspapers, magazines, books, radio, movies, and the Internet. The mass media, in its numerous forms, has become a fundamental part of contemporary life.

Historical Overview of Mass Communication in America

Throughout most of human history, speech and body language were the only forms of interpersonal communication. Communication changed with the development of writing in about 3,000 B.C. Information spread throughout North American colonies through letter carriers, postings in taverns, and via word-of-mouth. Rumors and gossip were considered primary methods for spreading the news. Mass communication dates from the invention of the printing press by Johannes Gutenberg in 1456, who created the means by which printed documents, most notably the Bible, could reach large numbers of people (Rodman, 2001). First utilized to propagate religious text, the printing press was soon used to distribute news, entertainment, and government missives. Newspapers made periodic appearances as early as the 1600s, in the very beginning of the colonial days (Compaine and Gomery, 2000). The first American magazines appeared in the 1740s. In 1791, Congress ratified the First Amendment, emphasizing the government's commitment to free speech and media freedom. Over the next 430 years, newspapers, books, and magazines were the primary methods by which information was presented for mass consumption (Rodman, 2001), until the advent of broadcast radio in the 1920s. In the past sixty years, media evolution has made rapid changes with the invention of the television, and more recently, the growth of cable television and creation of the Internet (Rodman, 2001). Today, the mass media, comprised of print media (books, newspapers, magazines), electronic media (television, radio, audio/video recording), and new media (computers and computer networking) is a dominant presence locally, nationally, and globally.

Mainstream Mass Media Versus Alternative Media

While the present study is concerned primarily with the "mainstream" mass media, it is important to recognize that there are "alternative" media sources which often highlight information ignored by mainstream sources. The mainstream mass media refers to media that are "easily, inexpensively, and simultaneously available to large segments of a population" (Surette, 1992: 10). Alternative media sources do not have the financial or political resources to reach the majority of the American public as compared to mainstream mass media sources. The mainstream mass media, which reaches the majority of the American population in terms of distribution numbers, have the greatest resources politically and financially. Alternative media sources were established in order to critique the mainstream mass media or to fill in the gaps created by narrow mainstream mass media agendas. Often, individuals seek out alternative sources through their own personal motivations, while mass media sources generally do not need to seek out consumers. The mainstream mass media sets the framework in which other media sources operate (Chomsky, 1997). Throughout the present study, the term "media" refers to the mainstream mass media.

Pervasiveness of Media Exposure in the United States

The mass media in the United States is comprised of 1,700 daily newspapers; 11,000 magazines; 9,000 radio stations; 1,000 television stations; 2,500 book publishers; and 7 movie studios (Bagdikian, 2000; Compaine and Gomery, 2000). According to Stempel and Hargrove (1996), in a 1995 survey of Americans, 70.3% were regular viewers of local TV news; 67.3% were regular viewers of network TV news; and 59.3% read a daily newspaper. In addition, 48.6% of the survey population listened regularly to radio news and 31.4% regularly read a newsmagazine. According to a Gallup poll in 1996, 78% of Americans claimed they get their news from nightly national television newscasts (DeLuca, 1999).

Television has become the dominant form of news communication. In 1950, only 9% of U.S. homes had a television set. Today, the average American household has two television sets, which are on for more than seven hours per day (Rodman, 2001). According to Graber (1980), the average American high school graduate spent more time in front of the TV than in the classroom. According to the National Association of Broadcasters (1995), the average person listens to the radio for over 22 hours per week.

The growth and development of the Internet and the World Wide Web has had a major impact on mass information dissemination especially in the past ten years. A survey conducted in 2001 by the UCLA Center for Communication Policy found that 72.3 percent of Americans had online access, a growth of over 5 percent from 2000 (Surette, 1998). According to the Census Bureau (2000), over 54 million American households or 51 percent had one or more computers, an increase of 9% in a little over a year. Internet and World

Wide Web use has increased dramatically in the past few years, making it the most significant communication tool ever devised (Greek, 1997).

Factors Influencing Media Information

Media Ownership

There is a growing concentration in media ownership (Bagdikian, 2000; Herman and Chomsky, 1988; Manoff and Schudson, 1986; Miller, 1996, 1998; Parenti, 1993). At the end of WWII, eighty percent of daily newspapers were independently owned. By 1989, eighty percent of daily newspapers were owned by corporate chains. In 1983, fifty corporations dominated the mass media and the largest media merger in history involved a $340 million transaction. By 1990, twenty-three corporations controlled most of the mass media. In 1997, just ten corporations dominated the mass media, and the Disney-ABC deal became the biggest merger in history at $19 billion. Today's mass media is virtually controlled by six firms, which are among the world's largest and most powerful corporations; General Electric, Viacom, Disney, Bertelsmann, Time Warner, and Murdoch's. In 2000, the AOL-Time Warner merger involved a $350 billion deal which was over 1,000 times greater than in 1983 (Bagdikian, 2000).

The ownership of the mass media by just six conglomerates means that a very powerful and prosperous few have control over influencing the American public. Media owners are driven by profits, most of which are derived from advertising dollars of other multi-national corporations. The voices of those opposed to the vested interests of media corporations are not likely to be heard (DeLuca, 1999). Former CBS president Frank Stanton stated, "Since we are advertiser supported we must take into account the general objective and desires of advertisers as a whole" (Parenti, 1993: 35). For example, Chrysler's advertising agency circulated a letter to magazines requiring them to submit articles for screening for possible offensive content to Chrysler (Glaser, 1997). The government appears indifferent to the immense and still growing power of major media corporations (Bagdikian, 2000). Citizen action groups and alternative media outlets lack the financial and political resources to match corporate funds. For the most part, the public isn't even aware of the political, social, and economic dangers of concentrated corporate control of the media.

Use of Authorities as Sources

There is a heavy demand for dramatic and sensational stories and the media must pick and choose which stories to present to the public. "If it bleeds, it leads", has become a leading media mantra. Journalists rely heavily on easily accessible and reliable sources for information which generally means using government officials. Media personnel are not likely to criticize governmental organizations out of fear they may deny access to information. Reliance on high-ranking officials is problematic for several reasons. Reliance on political officials leads to the acceptance and reaffirmation of traditional approaches to dealing with certain social problems. In addition, rather than provide accurate information, bureaucrats can use news exposure opportunities to promote themselves and the institution they represent, which leads the public to believe they are reliable and credible sources of information (Chermak, 1997).

Political and Social Impact of Mainstream Mass Media

The mass media is the platform by which a plethora of political matters are discussed and how most people learn about political issues and determine which are important (Perse, 2001). Because the public's exposure to the political process is limited (Kessel, 1975), information from the mass media may be the only contact with politics for an overwhelming majority of Americans (McCombs and Shaw, 1972). Political campaigns are often built around electronic media because they are a cost-effective way to gather support for policy positions (Graber, 1980; Skogan and Maxfield, 1981; Tunnell, 1992). One of the major uses of media in political campaigns is agenda setting.

Agenda-Setting

Agenda-setting refers to the power of the news media to direct our concerns toward certain issues (Perse, 2001). A large body of research supports the agenda-setting influence of the media (Berk, Brookman, and Lesser, 1977; Dearing and Rogers, 1996; Fisher, 1989; Gordon and Heath, 1981; Haskins and Miller, 1984; McCombs and Shaw, 1972; Pritchard, 1986). According to Cohen (1963), the media may not tell us exactly what to think, but they tell us what to think about. The similarity of programming across channels and in news reports due to the concentration of ownership and economy of scale have led to the proliferation of different venues of news drawing from the same sources (Perse, 2001). Repetition of certain issues, people, and events in conjunction with media consistency reinforces the public's understanding of what is important (Perse, 2001). McCombs and Shaw (1972), the original pioneers of the term "agenda-setting" found almost identical rank-order correlation between amount of news coverage of issues and the rank ordering of those same issues by a sample of individuals. Dearing and Rogers (1996) in a meta-analysis of 100 studies, found overwhelming support for the agenda-setting hypothesis. Funkhouser (1973a, 1973b) and MacKuen and Coombs (1981) found that the public's belief in the importance of events closely followed media coverage of events, and not real-world indicators. The media are a powerful force in establishing public opinion and in reducing the number of divergent opinions in society (Noelle-Neumann, 1991, 1993).

Conclusion

The mainstream mass media are a dominant presence in American society. On a daily basis, the American public is inundated with a vast array of media information from a variety of sources. The mass media not only provide the public with information but also interpret the information. In addition, the mass media have a great impact on socialization by providing a sense of collective norms and values. In the past several decades, the media has shifted from an investigative role to a more profit-driven role. The growing concentration of media ownership and the use of government officials as sources both have a huge impact on media content. Information presented to the public often reflects the interests of the powerful in our society. Although alternative media sources exist, they lack the financial and political resources to compete for exposure with mainstream mass media sources.

CHAPTER 3

The Mainstream Mass Media and Crime

INTRODUCTION

Many of our ideas about the world around us are gleaned, not through direct experience, but through exposure to the mass media. The news media play an important and primary role in the construction of social problems (Sacco, 1995). In many cases, the media distorts the facts and provides us with a simplistic and often erroneous view of reality. By creating a distorted, provincial, and/or false view of reality, the media are responsible for perpetuating myths that have dramatic, misdirected, and often dangerous consequences. In particular, the media has misconstrued the reality of crime in American society. According to Fishman (1978: 542) in an analysis of the social and media construction of crime waves, "the interplay between national elites and national media organizations may well have given rise to a number of social issues now widely accepted as fixtures in the recent American political scene". The media doesn't just report about crime; they are responsible for constructing a social reality of crime that has an enormous impact on public perceptions of crime and criminality (Surette, 1992; Barlow, 1991; Garofalo, 1981). Chapter Three presents a thorough overview of the literature concerning the mass media and street crime; discusses the limited media attention given to corporate crime; describes media reporting on the environment and environmental crime; and highlights the problems associated with media reporting of environmental risk, harm, and crime.

The Mass Media and Street Crime

Research into the relationship between the media and crime, although not a recent phenomenon, has gained a great deal of criminological attention in the past two decades (Barlow et al, 1995a; Lofquist, 1997). Researchers generally agree that crime as it is portrayed in the mass media is distorted and over sensationalized (Barlow et al, 1995a; Benedict 1992; Chermak, 1994; Kappeler et al, 1996), presents a misleading view of crime (Chermak, 1998; Fishman, 1978; Graber, 1980; Lotz, 1991; Marsh, 1989), and blurs the line between news and entertainment (Newman, 1990). Politicians, the public, and the media are preoccupied with violent crime and neglect other types of crime, in particular corporate crime (Kappeler et al, 1996). Furthermore, the media focuses a great deal of attention on crimes committed by young, male minorities while overplaying the prevalence of white, affluent victims. The media perpetuates the myth that most crime is interracial. The media makes us afraid of random violent crime by strangers and even though youth crime is on the decline, surveys indicate that an overwhelming number of Americans believe juveniles are committing more crimes than ever before. The picture of crime in America, as presented by the vast majority of media outlets, is of the violent stranger and as such, the most viable solutions are more police, more laws, and harsher sentencing practices. By limiting or excluding incidences of corporate crime from news coverage, the media plays a large role in shaping public opinion as to what constitutes crime (Garofalo, 1981; Hills, 1987; Marsh, 1989; Reiman, 1998). This distorted view of crime has an enormous impact on society. Fear of crime, in particular violent, individual crime, is on the rise even though the violent crime rate has been on the decline over the past two decades.

Most people have little direct experience with the types of crime presented in the media (Ericson et al, 1987; Graber, 1980; Hall et al, 1978; Stroman and Seltzer, 1985; Surette, 1992). Therefore, the public relies heavily on the media to supply them with crime news. Researchers emphasize that it is important for criminologists to challenge the media and analyze reporting biases (Wright et al, 1995). Criminologists have a great deal to offer the news media with respect to making news more representative and less distorted (Barak, 1994).

Sources of Information

Police and court officials provide relatively easy access to crime information. However, they also affect how crime is presented in the news. Reporters generally rely on authoritative sources for crime news (Berkowitz, 1987; Berkowitz and Beach, 1993; Brown et al, 1987; Chermak, 1995; Gans, 1979; Sigal, 1973). The media utilize the police and criminal justice officials as their primary source of information for a number of reasons. In order to provide the public with as much credible information as possible, the media need to gather information from reliable sources. In addition, due to time constraints, the media need easy and quick access to crime information. The police provide the media with seemingly credible and easy-to-access data (Lynch et al, 2000). The problem with relying on police information is that once again, certain crimes, moreover street crimes, are given more coverage than other types of crime and the police are able to promote their own interests and their own version of crime (Sherizen, 1978; Fishman, 1980; Hall et al, 1978; Ericson et al, 1987; Grabosky and Wilson, 1989). In addition, "the police role as the dominant gatekeeper means that crime news is often police news and that the advancement of a police perspective on crime and its solutions is facilitated" (Sacco, 1995: 146). Chermak (1997) found that in the majority of 1,900 crime, drug, and policy stories, police and court officials were utilized as sources. Criminologists and sociologists only accounted for 2% of sources in all crime stories and even less in drug stories (Chermak, 1997).

Focus on Individual Violent Crime and Neglect of Official Crime Data

Serious personal crime, most notably murder, is given high priority by the mass media (Cohen, 1975; Chermak, 1994, 1995; Ericson et al, 1991; Graber, 1980; Humphries, 1981; Sheley and Ashkins, 1981; Skogan and Maxfield, 1981) while white-collar crime and property crime are given very little attention (Chermak, 1994, 1995; Evans and Lundman, 1983; Graber, 1980; Jerin and Fields, 1995). A large amount of criminological literature supports the premise that there is an overrepresentation of violent individual crimes in the news media, especially when compared to proportions of such crimes indicated in the official crime data (Barlow et al, 1995a; Graber, 1980; Garofalo, 1981; Reiman, 1998; Sherizen, 1978; Skogan and Maxfield, 1981).

Barlow et al (1995a) found that 73% of the articles included in their sample of news magazines focused on violent crime whereas only 10% of crimes known to police involved such violence in that same year. According to Chiricos et al (1997) television and news stories about violent crime and juvenile violent crime increased more than 400% between June and November of 1993. However, while media and public attention to violent crime continued to escalate, the rates of such crime continued to decline (Chiricos et al, 1997).

Media accounts of crime not only exaggerate incidences of violent crime, they egregiously overstate the occurrence of individual crime (Garofalo, 1981; Graber, 1980; Schlesinger et al, 1991) and stranger crime (Chermak, 1994; Kappeler et al, 1996; Tunnell, 1992.) The reason for the overrepresentation of violent individual crime has a lot to do with the sensational and dramatic quality of such crimes (Sacco, 1995). Although these crimes are atypical, they provide the media with the opportunity to create dramatic stories with victims and villains.

Distortion of Victim and Offender Characteristics

A number of studies examining the nature of homicide reporting have found that the strongest predictor of reporting and attention was directly related to the number of victims killed during the incident (Chermak, 1998; Johnstone et al, 1994; Wilbanks, 1984). In other words, the more victims, the more coverage. Several studies show that minorities are overrepresented as offenders in news coverage of crime (Barlow et al, 1995a; Sheley and Ashkins 1981; Smith, 1984) and there is a growing emphasis on socially favored victims of crime (Benedict, 1992; Fishman, 1978; Graber, 1980). Barlow et al (1995a) found a significant bias against racial minorities in the news accounts of crime utilized in their study of Time magazine articles over a five-year period. While official data reported that white offenders were responsible for the majority of crimes committed in the years in question, over 74% of news reports on crime during the same time frame concerned minority offenders. Similarly, Entman (1990, 1992, 1994) found that defendants were most likely to be presented as African-American. Humphries (1981) found a disproportionate emphasis on the arresting of young minority males from lower class backgrounds in his study of news stories in the New York Post in the 1950s and 1960s. Johnstone, Hawkins, and Michener (1994) found that murders of minority victims were less likely to be reported while murders of women and children were more likely to be reported.

<u>Lack of Attention Given to Solutions or to the Wrong Solutions</u>

Most crime news articles focus on criminals and criminal events with little attention given to solutions to the problem (Barlow et al 1995a; Sherizen, 1978; Dussuyer, 1979; Graber, 1980). For example, in their study of 175 Time magazine articles, Barlow et al (1995a) found that 82% of the articles focused on crime and criminals and only a small percentage (17%) addressed larger criminal justice issues.

While lack of media attention to appropriate solutions is cause for concern, even more troubling is the attention given to solutions that have little or no positive support in the academic literature. Cavendar (1984) studied the media coverage of "Scared Straight", a program designed to bring troubled juveniles into contact with inmates in a New Jersey prison. The program was one of the most widely publicized media presentations of crime in the 1970s. Although evaluations of the program and similar "shock" programs failed to produce significant results in the criminological literature, the media nonetheless promoted the ideals of deterrence and retribution as primary punishment mechanisms for reducing criminal and delinquent behavior (Cavendar 1984).

Federal anti-crime agendas have prioritized criminalization and enforcement over social intervention since the early 1920s (Potter, 1998). Anti-crime legislation which focuses on getting tough and pointing the finger at individual responsibility continues to dominant the political and social agenda while there continues to be almost no mention of economic and political structures as root causes of crime (Barlow et al, 1995b). By presenting crime as largely the result of individual pathology, the media neglect to link crime with broader social forces (Humphries, 1981). When the public believes violent crime is so prevalent and that police are very successful in apprehending offenders, they will continue to support legislation and funding that calls for more police, more prisons, and more money for the criminal justice system (Surette, 1992).

<u>Creating Fear</u>

By promoting violent and individual crime, the media has the potential to elevate fear of crime or fear of certain types of crime. Williams and

Dickinson (1993) articulate that regular exposure to crime news has a direct impact on fear of victimization. While Sacco (1995) emphasizes that many consumers are skeptical of the news media, there remains a substantial number of people who believe what they watch and read. Heath and Gilbert (1996) suggest that some television viewing is correlated with fear of crime for some viewers. However, directly relating fear of crime to media exposure is difficult to uncover due to the complexities of the relationship between fear and media exposure. The type of programming, operationalization of fear, viewer demographics and beliefs, sense of justice, and level of fear prior to exposure all have an impact on study results. Therefore analyses of exposure to crime news and fear of crime is difficult to accurately determine. Heath (1984) found in a sample of phone interviews that reports of local crimes that were sensationalized or random were associated with higher levels of fear of crime. Similarly, Williams and Dickinson (1993) found that British news articles depicting more sensational aspects of crime appeared to promote fear of crime. Gordon and Heath (1981) found that fear of crime is related to the proportion of the newspaper devoted to crime. Liska and Baccaglini (1990) found that fear of crime was greater in middle aged white women. The researchers suggested that the elevated fear among this group was due to their overrepresentation as victims on television news shows. In reality, middle-aged white women are less likely to be victimized than young, minority males. Elevated levels of fear in women have devastating effects on women's feelings of independence and thwarts efforts to be powerful in a male dominated society.

In Support of the Powerful

The overrepresentation of violent individual crime and underrepresentation of corporate and other forms of crime in the news media has serious consequences. Several researchers emphasize that crime news supports the interests of the powerful in our society (Hall et al, 1978; Barlow et al 1995a) and diverts public attention away from the enormous impact and costs associated with crimes committed by the elite and powerful members of society (Wright et al 1995, Hills, 1987; Reiman, 1998). "Equating crime with violence, rather than recognizing it for what is most often is-the acquisition of property-distorts the property relations in capitalist society, which makes most crimes so conspicuously rational" (Barlow et al, 1995a: 10). In addition to failing to take into consideration the links between crime and unemployment, the news media rarely if ever suggests that macro-social conditions are the source of the crime problem (Barlow et al 1995b). Marxist media critics emphasize that the media has become the means by which the "haves" of society gain the willing support of the "have-nots" in order to maintain the status quo (Rodman, 2001). In other words, the mass media distract people from the "real" problems existing in society such as poverty, racism, sexism, and the like in order to emphasize the threats of individual and violent predators. Crime has never been abolished but the federal government has succeeded in expanding its capacity to police the nation through the identification of public enemies and the creation of new crimes (Potter, 1998).

In recent years, the media has focused its attention on a handful of corporate scandals, namely Enron and Martha Stewart. While the media can be credited with providing the public with information regarding such incidents, there has been a limited amount of critical dialogue concerning these types of corporate crimes. The focus in both cases has been on individual accountability and not on the corporate, economic, and social climate which often encourages such behavior.

The Mass Media and Corporate Crime

Several researchers have estimated that the costs of corporate crime in terms of direct financial costs to consumers exceeds $2 billion annually (Clinard and Yeager, 1980; Kappeler et al, 1996; Simon and Eitzen, 1993). Despite the enormous costs associated with corporate crime, the media generally ignores or underestimates the costs of corporate crime (Hills, 1987; Kappeler et al, 1996; Reiman, 1998). Additionally, a large number of studies have suggested that the human costs in terms of death and injuries due to corporate crime are greater than those associated with street crime (Bierne and Messerschmidt, 1991; Clinard and Yeager, 1980; Frank and Lynch, 1992; Kappeler et al, 1996; Michalowski, 1985; Reiman, 1998). The costs in terms of dollar amount and human injury/death due to corporate crime is enormous, yet there is very little attention directed toward this type of crime from the media, the public, politicians, and even within academia. Calavita and Pontell (1994) suggest that even if corporate crime is depicted as a threat in media reports, it is generally described as a threat to business and economic interests rather than consumer, employee, and environmental interests.

Criminologists who study white-collar crime in its various forms realize that it is much more complex and more difficult to reduce to numbers than street crime. Just as there is relatively little media coverage of corporate crime, a limited number of studies have examined representations of corporate crime in the media (Evans and Lundman, 1983; Lofquist, 1997; Lynch et al, 1989; Lynch et al, 2000; Morash and Hale, 1987; Randall, 1987; Randall and Lee-Sammons, 1988; Swigert and Farrell, 1980; Wright et al, 1995). The research indicates that reporters appear to have an inadequate and simplistic understanding of the complexity of corporate crime (Levi, 1994; Randall, 1987; Randall et al, 1988) and are unlikely to conceptualize corporate deviance as "crime" (Lynch et al, 1989; Wright et al, 1995). Evans and Lundman (1983) and Morash and Hale (1987) examined cases of non-violent corporate crime. News coverage in both cases was limited and accounts directed attention toward individual responsibilities or secondary causes rather than organizational malfeasance (Hills, 1987; Morash and Hale, 1987; Wright et al, 1995b).

Lofquist (1997) compared newspaper coverage of two widely reported crimes that occurred in Rochester, New York in 1994. The first case involved the disappearance of Kali Ann Poulton, a 4 year-old girl, while the second case centered on the collapse and flooding of a large salt mine owned by Azko Nobel Salt. The cases were similar in that they occurred in the same year, in the same area, and it was unclear as to whether they were actually accidents or crimes. Detailed analyses of news coverage of these events revealed that the media immediately depicted the missing child as a victim of stranger abduction, despite the fact that stranger abductions are rare cases. Family members or acquaintances are most likely to be responsible for child abductions. In the case of the mine collapse and subsequent flooding, despite overwhelming evidence of corporate negligence, the media described the event as an "accident". Lofquist (1997: 256) concludes that the media is responsible for creating and reproducing hegemonic understandings of events. In other words, the media chooses to fill in the gaps in ways which protects the dominant social structure and points the finger at individual actors as responsible for such criminal events. In the case of Kali Ann Poulton, the media created a social reality that suggests that our children are in grave danger of pathological strangers rather than calling into question the dangers of poverty, illiteracy, poor education, poor health care, and the like (Lofquist, 1997). In the case of corporate negligence, Lofquist highlights that "organizational wrongdoing is obscured; the weakness of regulation and of media scrutiny limits the likelihood of 'naming and blaming' and allows a vocabulary of 'accident' to prevail" (258).

Research on media coverage of corporate *violence* is even more limited. According to Lynch et al. (1989) the American media is reluctant to socially construct corporate violence as crime. Wright et al (1995: 22) stresses that "how the media constructs corporate violence can affect whether it will be conceptualized and treated as a crime". Swigert and Farrell (1980) examined newspaper coverage of corporate violence in reference to the Ford Motor Company's Pinto scandal. Ford's failure to recall the Pinto resulted in numerous injuries and deaths to consumers. Reporters initially portrayed the cases as indicative of accidents and not corporate violence. News coverage gained momentum when it was discovered that Ford officials were aware of the mechanical defect and refused to recall the Pinto (Dowie 1977). Swigert and Farrell (1980) contend that media attention to the case contributed to Ford's eventual indictment and prosecution on charges of reckless homicide.

Even in cases in which evidence of corporate violence is clear and convincing, the media still has difficulty linking such behavior with crime. Wright et al (1995) analyzed newspaper coverage of a fire at the Imperial Food Products plant in North Carolina. The fire resulted in 25 deaths and over 55 injuries. It was widely reported that the exit doors had been locked or barricaded by the owner; there was no plant-wide working sprinkler system; no windows; too few exits; and the plant had never been inspected by OSHA (Occupational Safety and Health Administration). Wright et al (1995) reasoned that the case provided a unique opportunity to study the media's reactions to corporate violence. The evidence of corporate malfeasance was strong, the physical harm severe, and the case ended in charges of manslaughter. Often, in cases of corporate violence, a clear individual offender is difficult to find (Clinard and Yeager, 1980). Wright et al (1995) conducted a content analysis of 10 major city newspapers. Nine of the ten papers covered the fire but new coverage dwindled substantially over the days following the fire. The news coverage focused mainly on the enormous death and physical harm caused by the fire and the suffering and damage to the community. But although the incident was immediately perceived to be an act of corporate violence, the media did little to link such actions with crime (Wright et al, 1995). The deaths were not depicted as homicides nor was the possibility of prosecution raised until after the government indicated its intent to prosecute. Even when the case officially became a crime, news coverage still did not depict the actions as criminal. "Instead of a potential criminal offense, the news reports socially constructed the worker deaths as a breakdown in government safety regulation" (Wright et al, 1995: 32). Consequently, the coverage did not transform the public reality of corporate violence as crime. Furthermore, the limited coverage of the manslaughter convictions did little to educate the public or produce deterrent effects (Wright et al, 1995: 32). Lynch, Stretesky and Hammond (2000) argue that crime news is constructed not only by what is said about corporate crime, but by what is left out. That is, the public image of crime is shaped by the nonreporting of corporate crime. Underreporting the extent of corporate crime and, at the same time overreporting on crimes the public fears the most, both shape the fear of crime.

The Mass Media and the Environment

For the most part, the media's interest in the environment and related issues is cyclical (Gaber, 2000). There is a great deal of media coverage during environmental disasters and industrial catastrophes but the attention quickly fades until the next crisis occurs (Anderson and Gaber, 1993). Consequently, there is almost no media dialogue concerning the true causes of such environmental devastation. Furthermore, mass media's focus on spectacular events prevents sustained coverage of the more serious environmental problems facing our society (DeLuca, 1999).

The role of the mass media in the history of environmentalism has not received a great deal of attention (Neuzil and Kovarik, 1996). Ponder (1986) examined the role of the media in environmental dialogue during the Progressive Era and suggested that the media were active in calling for environmental reform on the federal level. According to Neuzil and Kovarik (1996), "from the muckrakers' work in the public health reform movements to scientific and political fights to conserve western lands and resources, journalists participated in many environmental controversies of their era" (1996: xxi).

The influence and popularity of television in the 1960s had a great impact on environmental awareness (Neuzil and Kovarik, 1996).

Although the mainstream media had little or nothing to do with environmental legislation in the 1970s, research suggests that alternative media outlets and activists had a great deal of impact on the creation and implementation of federal environmental policy during that time (Neuzil and Kovarik, 1996). Prior to the creation and enactment of federal environmental legislation in the 1970s, specialized environmental publications and professional interest groups were calling for political involvement in environmental issues at the national level. The mainstream media gave attention to environmentalism and environmental policy *after* the legislation was already in place (Strodhoff, Hawkins, and Schoenfeld, 1985).

Critical media attention toward corporations in the early 1970s, spurred by the consumer movement and actions of Ralph Nader, angered and outraged corporate leaders. In addition to pouring millions into elaborate PR campaigns and lobbying efforts, corporations launched a savage campaign against the media. Corporate leaders attacked the media, suggesting that the media had a significant bias against business. In 1980, corporate leaders were successful in electing a national administration dedicated to wiping out a half century of social legislation and regulation of business (Bagdikian, 2000). Today, with ownership of the mass media in the hands of just six corporations, media reporting is heavily weighed in favor of corporate values.

In the past few decades, the public has been inundated with specialty environmental magazines, books, and cable television shows and channels. According to American Opinion Research, Inc. (1993) by 1993, more than two-thirds of the nation's medium and large newspapers had reporters specializing in covering issues involving the environment. Despite the growth in environmental awareness across the media, political, and public realms, most media information focuses on what individuals can do to save the environment. The mass media has aided corporations and the politicians in creating a consumer culture that advocates individual responsibility for protecting the environment. According to a Fairness and Accuracy in Reporting survey which analyzed source attributions in news articles relating to environmental issues, fifty percent of all quotes come from government officials (McDonald, 1993). The next largest percent of environmental quotes came from industry and the lowest percentage, 4%, from environmental groups. Consequently, environmental issues, as depicted in the media, are presented in government and corporate terms. The media then rarely questions the structural components that have led to environmental harm.

Radical Environmental Groups and the Media

Environmental organizations have utilized a wide range of tactics to gain media attention and publicity for environmental issues. Greenpeace was one of the first environmental groups to recognize the power of the mass media to publicize their efforts. Since 1971, environmental activists have performed thousands of "image events" in support of environmental issues including chaining themselves to whaling harpoons, plugging waste discharge pipes, and forming human blockades to stop trucks from transporting hazardous waste (DeLuca, 1999). Members of Earth First! have sat in trees, blockaded roads, and chained themselves to logging equipment. In many ways, these environmental activists have been successful. There is a ban on commercial whaling and ocean dumping of nuclear waste and activists have successfully blocked the placing of several garbage and hazardous waste incinerators. Environmental groups have gained more public visibility and public support for environment issues.

Despite the number of successes achieved by these radical groups, they have a very uneasy relationship with the media and more often than not, these radical environmental groups are depicted as crazy, deviant, and "disturbers of order" (Parenti, 1993). Corporations have filed lawsuits against many environmental activists. Many activists have been the victims of vandalism, death threats, and serious violence. For example, on September 17th, 1998, David Chain, an Earth First! member was crushed to death by a redwood when an angry Pacific Lumber logger continued to fell trees despite the presence of protesters (Goodell, 1999). Corporate leaders, politicians, and the FBI have labeled many activists as terrorists, even activists who themselves have been victims of threats (DeLuca, 1999). According to Lois Gibbs, founder of the Love Canal Home Owners Association and later, founder of the Center for Health, Environment, and Justice, states that "people have been followed by private detectives, had their homes broken into. I'd say 40 percent of people protesting toxic waste sites and incinerators around the country have been intimidated (Helvarg, 1994: 651).

Radical environmental groups maintain that confrontational efforts and orchestrated image events are the only major ways to achieve massive publicity and support. Elected officials, corporate leaders, and corporations all enjoy enormous advantages over environmental groups in terms of access to the media and control of their image, which "is due in no small measure to the fact that media themselves are giant corporations with a vested interest in the status quo" (DeLuca, 1999: 20). Image events are intended not only to bring attention to a particular imminent environmental issue; they are intended to contest the hegemonic discourse of industrialism that dominates our society (DeLuca, 1999). Image events though are rarely recognized as working for social structural change. "News media's emphasis on the new, its quest for the novel, forces groups to perform even more outrageous events in order to get coverage" (DeLuca, 1999: 92). Radical environmental groups are in a difficult position. In order to get public attention, they must rely on the media to cover environmental issues. The media will only give radical environmental groups attention when the story is exciting and dramatic. The confrontational tactics utilized by radical activists often come across as crazy and desperate.

Mainstream environmental groups are a great source of animosity for radical environmental groups. Mainstream groups appear to be working for the environment in socially and politically appropriate channels and therefore come across as diplomatic and responsible engineers of environmental protection. Despite their public image as supporters of the environment, most mainstream groups are aligned with corporations, industry, and government. It's almost impossible to tell them apart. For example, Jay Hair, former president of the National Wildlife Federation now does public relations for Plum Creek Timber (Cockburn, 1997). John Sawhill, president of the Nature Conservancy, appears in General Motors ads which tout the shared goal of "safeguarding the environment without destroying jobs or businesses" (DeLuca, 1999). Mainstream groups, as allies of government and industry, often adopt anti-environmental initiatives (Cockburn, 1995; Dowie, 1995; Sale, 1993). They advocate and promote market solutions to environmental problems and rarely, if ever, challenge the industrial exploitation of nature (DeLuca, 1999). The government, corporate leaders, and the media frame radical environmental groups in negative terms because they fear the disruption of their power and privilege (Gitlin, 1980). The media understands environmental issues, groups, and disasters through the discourse of industrialism (DeLuca, 1999). Environmental columnist Edward Flatteau stated that "there are exceptions, but publishers are basically hostile to environmental protection. It's a threat to their business. Their economic lifeblood comes from advertising revenues and that means conspicuous consumption" (Jacobson, 1998: 48).

Media Reporting of Environmental Crime

There are only a handful of studies that have examined media coverage of corporate crime and even fewer studies have examined media coverage of environmental crime. Lynch, Nalla, and Miller (1989) analyzed media coverage of the Union Carbide lethal gas leak in Bhopal, India, which resulted in the immediate deaths of over 2,000 people. The authors compared articles and pictorial representations of the event as depicted in American and Indian magazines. American magazines portrayed the event as an "accident" or as a disaster and labeled Union Carbide as a victim. Conversely, Indian magazines labeled the event as a crime and portrayed Union Carbide as the negligent offender. Similarly, Lynch, Stretesky, and Hammond (2000) emphasize that most environmental problems and disasters (i.e. pollution, hazardous waste dumping/siting) are described in the news media as accidents. In addition, the authors suggest that it is common to depict environmental pollution as the "price we pay for technology" (Lynch et al, 2000: 115). Lynch et al (2000) found that only eight (1.5%) of 544 cases of chemical crimes in Tampa were actually reported in the Tampa Tribune. Of the eight articles discussing chemical crimes in Tampa, two indicated that the crimes were accidents and the other six articles suggested that poor individual decision-making was the cause of the chemical incidents. Furthermore, while there were only forty-seven homicides in Tampa in 1995, there were eighty-eight articles on these particular homicides and 4,089 articles concerning homicide in general. Overall, the study found that there was no discussion of corporate negligence in news media coverage of environmental crime in Tampa. The authors conclude that more research is necessary in order to determine the prevalence across news media outlets of neglecting and ignoring corporate crime, in particular.

Problems with Reporting Environmental Risk, Harm, and Crime

The mass media rarely unites issues of the environment, crime, and public health. One reason the media often avoids presenting information regarding environment risk and harm has to do with the complexity of the information. Environmental risk, harm, and crime are complex, multi-faceted, and deeply rooted in our political economy. Therefore, risks from dramatic or sensational causes of injury, illness, or death such as accidents, homicides, and natural disasters tend to be greatly overestimated while risks from environmental toxins and pollutants tend to be greatly underestimated (Lichtenstein et al, 1978). News media coverage of dramatic and sensational examples contributes to the difficulties of obtaining a proper perspective on environmental risks (Combs and Slovic, 1978). Psychological research demonstrates that people's beliefs change slowly and are extraordinarily persistent even in the face of contrary evidence (Nisbett and Ross, 1980). Consequently, public opinion is difficult to change. With constant media attention to random violent encounters and lack of exposure to the extent and severity of environmental harms, it is unlikely the public will regard environmental risks as serious.

In addition to problems encountered in reporting the complexity of environmental harm, other difficulties further impede media coverage of environmental issues. Reporters often rely on journalistic precedence when reporting information. The lack of precedence and lack of understanding of environmental harms has an impact on reporting. Environmental issues are not black and white and sometimes there are no clear victims and offenders. And since the government focuses very little attention on environmental issues, the media often regards such issues as less important and not newsworthy (Simon, 2000). Environmental risk, harm, crime, and justice are considered too difficult, too time-consuming, and too expensive to cover (DeLuca, 1999). According to Tom Winship, former editor of the Boston Globe, "there isn't a 'Stop the presses!' kind of development on the environmental story everyday. This is not event coverage. We need to persuade the media to cover the environmental story consistently. Sure, it's a slow story, but they've got to change their attitudes about what makes a story" (Hertsgaard, 1990: 16-17).

Conclusion

While "street crime" is given more than its fair share of media, political, and enforcement attention, "white collar" crime is generally ignored unless the consequences of such corporate actions results in several immediate deaths, affects hundreds or even thousands of lives, and costs several hundred millions of dollars (i.e. Enron). Even then, media attention is terminal. Headlines and leading news reports favor the isolated violent encounter. Although both "street crime" and "white collar" crime involve violence, victims, offenders, and injury, "street crime" is more sensational and simplistic and therefore, more appealing for copy than the often misunderstood and more injurious "white collar" variety. The media, our government and our justice personnel convince us that street crime is rampant and that we are all potential victims; worst case scenarios dominant our thinking and appear to be the norm. Consequently, voters are affected by this slanted portrayal. Over the past twenty years, we've become extremely adept at waging war against street criminals. Each year we build more and more jails and prisons. The war on street crime has diverted our attention away from the more serious problem of white-collar crime and corporate crime, despite recent headlines devoted to coverage of Enron and Martha Stewart. As long as we conceptualize street crime as the major criminal threat to society we will continue to ignore far more deadly, costly, and destructive crimes of corporate America. This cultural image of our crime problems is fed by the media, politicians, and crime specialists who emphasize the growing epidemic of the war on drugs, school violence, workplace violence, terrorism, and the like. In essence, murder by gun, knife, or other weapon is considered horrendous while murder by unsafe working conditions, pollution, and defective products is accidental and therefore, not as problematic or deserving of public attention. To compound the problem, many of the individuals who commit white collar offenses are the very same individuals who have the power, resources, and influence to shape laws and determine where much of our federal and state money goes. White-collar crime doesn't fit prevalent stereotypes of "real" crime hence it is not given as much attention by the media, politicians, the public or academics.

Media attention to environmental crime and its impact on the environment and human health is lacking. Given the importance of the media in creating public awareness and garnering attention for certain social problems, it is essential for researchers to examine media coverage of environmental crimes. There is no single type of "environmental crime"; consequently, research focusing on environmental crime must selectively examine a more narrow range or particular type of environmental offense and offender. Chapter Four introduces and describes one of the most polluting industries in the United States: the Petroleum Refining Industry.

CHAPTER 4

The Petroleum Refining Industry

INTRODUCTION

The petroleum refining industry is one of the leading manufacturing industries in the United States. Oil and natural gas are our biggest source of energy in the United States (65%) (American Petroleum Institute, 2004). Our nation uses two times more petroleum than natural gas or coal and four times more than nuclear power or renewable energy (Department of Energy, 2004). Oil is a valuable commodity and few individuals realize just how many products come from oil including gasoline, heating oil, plastics, diesel fuel, jet fuel, rubber, nylon, kerosene, tires, asphalt and even crayons. Chapter Four describes the current status of the petroleum refining industry; emphasizes the environmental and human health hazards associated with the industry; discusses the industry's environmental compliance history; and presents the literature related to petroleum refining industry violations. Only one study to date has examined media coverage of petroleum refining industry violations.

Current Status of the Petroleum Refining Industry

The United States is currently one of the largest producers and consumers of crude oil in the entire world. According to the Department of Energy (1998), in 1995, the United States was responsible for 23% of world refinery production. Almost fifty percent of the oil we consume is produced in the United States (American Petroleum Institute, 2004). Americans continue to consume about two-thirds of the world's oil production. Domestic production has declined but demand continues to soar. In the early 1980s, our country had a record high of 324 refineries and produced approximately 18.6 million barrels of oil per day. Today, the number of American oil refineries has decreased due to changes in oil prices, a shift to alternate fuel uses, and a focus on conservation (Envirotools, 2004).

Oil is a finite resource and accordingly production will eventually rise to a peak, which can never be surpassed. Once the peak has been passed, production will decline until oil resources are depleted. This peak effect is known as the Hubbert Peak (EcoSystems, 2004). According to a study conducted by Dr. C.J. Campbell on behalf of Petroconsultants (the most comprehensive database on oil resources outside of continental North America), world oil reached the midpoint of oil depletion in 1999. The study cautions that we are not running out of oil but we are running out of low cost, easy access oil that has fueled the economic development of the twentieth century (EcoSystems, 2004). The only companies a significant way from their midpoints or Hubbert Peaks, are the major Middle Eastern oil producers. Consequently, the likelihood of a global crisis similar to the oil crisis of 1973 is eminent.

The United States has found it increasingly difficult to balance diplomatic relations with Arab oil-producing nations while continuing to aid Israel (Foner and Garrarty, 1991). The petroleum refining industry faces some economic pressures with respect to increased costs of labor, compliance with new safety and environmental regulations, and the closing of small refineries. However, despite these pressures, total refinery output has remained steady and demand is increasing (EPA, 1995a).

The petroleum refining industry is comprised of a very small number of companies and facilities. According to the Census Bureau (1997) there are approximately 242 petroleum refineries in the United States. The EPA, which only includes larger facilities, estimates that there are 150 petroleum-refining facilities in the United States (EPA, 2004c). Table 2 presents the top U.S. companies with petroleum refining operations. While smaller refineries comprise half of the total number of refineries, they only produce approximately 14% of the total crude distillation capacity (EPA, 1995a). Most petroleum is refined and produced by large, integrated companies. The majority of facilities are located near crude oil sources which are concentrated along the Gulf Coast and in heavily industrialized areas on the east and west coasts. According to the Department of Energy (1998), 78% of the crude oil distillation capacity is located in just ten states. According to the 2001 Annual Survey of Manufacturers (Census Bureau, 2001), 101, 452 people are employed by the petroleum refining industry. In 2001, the value of shipment products sold by the refining industry totaled over $219 billion, which was approximately 5.5% of the entire U.S. manufacturing sector.

Table 2: Top U.S. Petroleum Companies 2002

Exxon-Mobil
BP
Royal Dutch/Shell
Chevron Texaco
TotalFinaElf
Conoco Phillips

Environmental Hazards Associated with the Petroleum Refining Industry

There are numerous air, water, and soil hazards associated with the petroleum refining industry and their processing methods. According to the Natural Resources Defense Council (2001), the petroleum refining industry is one of the major sources of pollution in the United States. The petroleum refining industry is the largest industrial source of volatile organic compounds; the second largest industrial source of sulfur dioxide; and the third largest industrial source of nitrogen oxides. Air pollutants include BTEX compounds (benzene, toluene, ethylbenzene, and xylene); carbon monoxide; hydrogen sulfide; sulfur dioxide; and methane (Envirotools, 2004). Air emissions are the result of equipment malfunctions, combustion processes, and transportation errors. Water pollutants contaminate the ground and surface water. Several refineries use deep-injection wells for disposal of wastewater. In many cases, this wastewater ends up polluting aquifers and groundwater. Soil pollution is generally the result of oil spills and landfill usage.

Air, water, and soil pollutants generated by the petroleum refining industry are directly related to a wide range of human health and environmental problems. Many of these toxic and hazardous air, water, and soil pollutants are known cancer-causing agents and are also responsible for liver damage and cardiovascular impairment. Human health consequences of exposure to petroleum refinery air pollutants also include gastrointestinal toxicity, kidney damage, blood disorders, reproductive and developmental toxicity, pulmonary disorders, polyneuropathy, cataracts, and anemia (EPA, 1995b). Benzene exposure is associated with aplastic anemia, multiple myeloma, lymphomas, pancytopenia, chromosomal breakage, and weakening of bone marrow (EPA, 1995b). In addition to causing a plethora of human health problems, exposure to pollutants generated by petroleum refineries causes a great deal of worry and fear among residents living near petroleum refining operations.

The decline in domestic crude oil output over the past decade has led to the demand for opening up additional areas for exploration and production. There is a great deal of controversy surrounding oil exploration in the Arctic National Wildlife Refuge in Alaska. According to a National Academy of Sciences report (2003), since oil was discovered, the environment has been substantially damaged due to refining operations. The future of the Arctic National Wildlife Refuge is in jeopardy. Oil industries spend millions lobbying legislators for reducing environmental standards and opening up additional areas for oil exploration. From 1992 to 1996, auto and oil industries gave more than $56 million in campaign contributions (U.S. PIRG, 1999). In 1998, auto and oil industries spent more than $90.9 million on lobby expenditures with Mobil, Exxon, and ARCO leading the way (U.S. PIRG, 1999). In addition, member of Congress who supported bills to overturn EPA air emissions standards received 76% more campaign contributions than members of Congress who did not support such legislation (U.S. PIRG, 1999).

Environmental Compliance

According to the EPA (1995a), the petroleum refining industry has a larger proportion of facilities in violation and with enforcement actions than any other industrial sector. The EPA's Petroleum Refining Compliance History analysis, which reviewed industry enforcement and compliance from August 1990 to August 1995, also found the following:

• Almost all facilities were inspected from 1990-1995 and on average, every three months.

• Facilities with one or more enforcement actions over the five-year period had, on average, eight enforcement actions brought against them.

• Of all the industrial sectors, the petroleum refining industry was the most frequently inspected.

• The rate of enforcement actions per inspection for the petroleum refining industry is high and has changed little over the past year.

• Clean Air Act violations were the most common.

According to the EPA's Toxic Release Inventory, which contains information on toxic chemical releases and other waste management activities, the petroleum refining industry released and transferred over 480 million pounds of pollutants in 1993 (EPA, 1995a). The petroleum refining industry is far above average in its pollutant releases and transfers per facility when compared to other industry facilities (EPA, 1995a). In 1993, seventy-five percent of the total poundage of releases involved air releases while twenty-five percent involved water releases. The petroleum refining industry was responsible for the release or transfer of over 100 different chemicals. Table 3 presents TRI information from 1993 for the petroleum refining industry.

Table 3: Petroleum Refining Industry TRI information, 1993

Percent of total pounds of TRI releases/transfers by all manufacturers	11%
Mean amount of pollutants released per facility	404,000 pounds (3.4 times more facility releases than other industries)
Mean amount of pollutants transferred per facility	2,626,000 pounds (13 times more facility transfers than other industries)

Environmental regulations have had a tremendous impact on the operations of the petroleum refining industry. Refineries have been forced to invest in upgrading their refining processes to reduce emissions. The refining industry has spent billions on complying with environmental regulations (Lichtblau, 1992). However, according to the Natural Resources Defense Council (2001), environmental laws and regulations do not stand in the way of expanding American oil refining capacity. American Petroleum Institute data indicates that oil refineries spend approximately one penny per gallon on clean air controls. Although the costs of complying with environmental laws have escalated in the past two decades, profitability has also been increasing. Oil companies are posting record profits (Natural Resources Defense Council, 2001). Joint ventures, mergers, and mega-mergers have allowed oil companies to reduce their costs by sharing operations and assets with other companies (Department of Energy, 2003). "Pollution abatement operating costs have been and continue to be a small part of overall operating costs" and play a small role in the deterioration of cash margins in U.S. refining and marketing (Department of Energy, 1997).

Petroleum Industry Violations

Only a few studies to date have examined petroleum refining industry violations (Randall and DeFillippi, 1987; Lynch, Stretesky, and Burns, 2004a, 2004b). Recently, Lynch, Stretesky, and Burns (2004a) examined whether petroleum refineries that violated environmental laws in Black, Hispanic, and low-income areas were more likely to receive smaller fines than refineries in White and more affluent communities. The authors found that "Black and low-income communities appear to receive less protection (via the deterrence goal of monetary penalties) from the EPA than areas with high concentration of White and high-income residents" (Lynch et al, 2004a: 436-437). The mean penalty for noncompliance in Black census tracts ($108,563) was much lower than in White census tracts ($341,590) and the mean penalty for noncompliance in low income census tracts ($259,784) was lower than in high income census tracts ($334,267). In a similar study examining petroleum refinery violations from 2001-2003, the authors found that refineries in Hispanic and low income zip codes received lower penalties than refineries located in non-Hispanic and more affluent zip codes (Lynch, Stretesky, and

Burns, 2004b). The authors conclude that penalty disparities are not the result of the seriousness of the violation, number of past violations, facility inspection history, facility production or EPA region but are the result of unequal protection of environmental laws for low income and minority communities (Lynch, Stretesky, and Burns, 2004a).

Media Coverage of Petroleum Refining Industry Violations

According to Randall and DeFillippi (1987), the media virtually ignored the oil industry prior to the early 1970s. However, following the oil embargo in 1973, the media and thus the public began to scrutinize the oil industry with more fervor than ever before. By the end of the 1970s, the oil industry had been accused of direct involvement in several incidents of illegal and unethical practices. Industry leaders angrily protested that the media had an anti-business slant. Leading corporate crime researchers, Clinard and Yeager (1980) stated that "the history of the oil industry has been characterized by the oligopolistic domination of the industry by a few massive corporations able to cooperate in controlling worldwide supplies and their distribution and thus to influence prices in a noncompetitive manner and a tendency for the federal government to defer to the power and interests of the industry".

Studies examining media coverage of petroleum refining industry violations are virtually nonexistent. Randall and DeFillippi (1987) examined patterns of media coverage of the 25 largest American oil firms from the late 1970s. The authors hypothesized that corporations with greater net sales, more frequent violations of law, and more serious offenses would receive greater media attention than corporations with lesser net sales, few law violations, and less serious offenses. Data were drawn from the Clinard-Yeager dataset for 1975-76, Moody's Industrial Manual for 1975, and news indexes for 1975-1976 (Wall Street Journal, Television News Index, and Reader's Guide to Periodic Literature). Randall and DeFillippi's (1987) content analysis revealed that the media attention was greater based primarily on the seriousness of the offense rather than the net sales of the firm or the frequency of offenses. The authors concluded that there was a "systematic media bias toward oversampling the most serious and undersampling the least serious oil firm violations" (40). Their results are not surprising considering the media's tendency to focus on the most serious violations of law across administrative, civil, and criminal categories.

Conclusion

Randall and DeFillippi (1987) offer one of the first and only studies of media coverage of oil company misconduct. For the most part though, their study is descriptive and offers little insight into what factors, other than perceived seriousness of the offense, result in greater media coverage. Furthermore, their data is drawn from oil industry and news source information from the late 1970s, over 25 years ago. In the past twenty-five years, there has been almost no academic inquiry into media coverage of the petroleum industry or on coverage of industry violations. The present study examines media coverage of federal petroleum refining industry violations in addition to examining the nature and distribution of this type of environmental crime. Chapter Five presents the data collected and methods utilized in the present study.

CHAPTER 5

Data and Methods

INTRODUCTION

The purpose of the present study is three-fold: (1) to determine the nature and distribution of petroleum refining industry violations; (2) to examine media coverage of petroleum refining industry violations and enforcement actions and determine whether media reporting is influenced by any specific case characteristics; and (3) to determine the impact of legal and extra-legal factors on fine amounts meted out to petroleum refineries found guilty of violating environmental protection statutes.

In order to accomplish these goals, data on media coverage of petroleum refinery violations, petroleum refinery violations, and community characteristics of areas where violative petroleum refineries were located were collected. News articles from twenty-five leading American newspapers were employed as the source for media reporting data. Data on petroleum refinery violations and area characteristics were collected from the Environmental Protection Agency (EPA). Descriptive statistics, content analysis and multiple regression were utilized to analyze the data.

Research Questions

To facilitate investigation of the issues described above, a series of research questions were devised. These questions are as follows:

1. What is the nature and distribution of environmental crime as indicated by federal petroleum refining violations?

2. What is the nature and distribution of mainstream news media reporting of federal petroleum refining violations?

3. Which factors lead to greater news media coverage of petroleum refining violations?

4. Are petroleum refining industry penalty assessment decisions affected by the racial and socioeconomic composition of the communities surrounding the violating facility? Do other factors influence penalty assessment decisions?

Data

Data for the present study were collected for the years 2001-2002 from the Environmental Protection Agency and for the years 1997-2003 from the LexisNexis database. Cases in the EPA database reflect cases settled or initiated in 2001-2002, consequently, some cases were initiated as early as 1997. The first step in this research was to identify all environmental violations by petroleum refineries using data from the EPA. Once identified, each case was searched in the LexisNexis data base in order to locate newspaper articles that reported on known oil refinery violations. The following sections describe the specific databases and the variables drawn from each case and article.

Environmental Protection Agency

Established in 1970, the Environmental Protection Agency (EPA) is the federal agency responsible for protecting human health and the environment by overseeing, developing and enforcing environmental policies and regulations. The EPA has an operating budget of over $7.6 billion and employs over 17,600 employees, making it the largest federal regulatory agency in the United States. Of particular interest in the current research is the EPA office of Compliance and Enforcement Assurance (OECA) which is responsible for compliance assistance, monitoring, incentives, and auditing as well as civil and clean-up enforcement. The goal of OECA is to maximize compliance and reduce threats to public health and the environment through coordinated efforts with state and local governmental agencies. EPA compliance and enforcement efforts are managed by a number of various sub-agencies including the Federal Facilities Enforcement Office (FFEO); Office of Compliance (OC); Office of Criminal Enforcement, Forensics, and Training (OCEFT); Office of

Environmental Justice; Office of Federal Activities; Office of Planning, Policy Analysis and Communication (OPPAC); Office of Regulatory Enforcement, and the Office of Site Remediation Enforcement.

Compliance and Enforcement Data: ECHO

The data used in the present study were collected from the EPA's Enforcement and Compliance History Online (ECHO) system. ECHO supplies compliance and enforcement data for over 800,000 regulated facilities nationwide and includes information pertaining to permits, inspections, violations, enforcement actions, and penalty information covering the past two years. ECHO data includes violations of the Clean Air Act (CAA) for stationary sources, the Clean Water Act (CWA) for facilities with direct discharge permits (under the National Pollutant Discharge Elimination System, NPDES), and the Resource Conservation and Recovery Act (RCRA) which includes information on generators/handlers of hazardous waste. Data on four key enforcement actions can be found in ECHO data (EPA, 2004a):

- The number of EPA inspections, and voluntary compliance or self-reported violation and pollution emission reports;

- The number and types of violations (noncompliance);

- The occurrence of a government enforcement action to address violations; and

- Penalties associated with enforcement actions.

The data for the present study were accessed through the EPA enforcement case search, which provides access to federal civil enforcement data tracked by the Integrated Compliance Information System (ICIS). ICIS is a multi-statute case activity tracking and management system for EPA administrative and civil judicial enforcement cases. Case information is supplied and updated by case attorneys in the EPA's Office of Regional Counsel and the Headquarters Office of Regulatory Enforcement (EPA, 2004b).

Data included in the present study were selected based on Standard Industrial Classification (SIC) system code 2911, which includes facilities in the petroleum refining industry engaged in producing gasoline, kerosene, distillate fuel oils, residual fuel oils, and lubricants through fractional or straight distillation of crude oil, redistillation of unfinished petroleum derivatives, or cracking or other processes (OSHA, 2004). In addition, cases were included in the present study if they were initiated or concluded between January 1997 and January 2003. Bounding the time period yielded 162 cases.

Using ECHO, a detailed case report summary of enforcement activity, and a detailed facility report was produced for each facility. Information was gathered on the following variables.

Company Information: Data were gathered on company name, address, city, state, zip code, and latitude and longitude. These data were used to identify each facility, and to allow each facility to be associated with Census data.

Case Type: Enforcement cases were either administrative or judicial. It is important to distinguish between cases resolved by judicial or administrative means. For example, it is possible that judicial cases are more likely to receive higher penalty assessments due to the fact they were not resolved without court intervention. In contrast, administrative decisions are typically rendered when the corporation and the EPA reach an informal agreement concerning an appropriate solution to the alleged violation. Thus, it could be hypothesized that because judicial cases are more likely to receive higher fines and compliance costs, that they are also more likely to receive media coverage, especially if the civil action is costly.

Voluntary Disclosure: For each case, the EPA indicates whether or not the case was the result of a facility self-disclosure or an EPA enforcement action. Theoretically, it could be hypothesized that the EPA is likely to be more lenient with facilities that self report law violations, and that it is more likely to require compliance without assessing a penalty. In addition, self-disclosed cases that result in financial penalties are more than likely to receive a lower penalty assessment than cases discovered through the EPA inspection process.

Multi-media: The EPA records whether or not the facility was in violation of more than one environmental statute. If the facility was in violation of more than one environmental statute during an inspection, the case is considered to be a multi-media case. Penalties should be higher in cases involving more than one environmental statute violation.

Case Status: Cases were either coded as closed/concluded or in process/other. No penalty amount can be determined for cases without enforcement outcomes, consequently some data will be coded as missing. Media reports, however, may be available for ongoing cases.

Case Outcome: Case outcomes fall under one of six categories; final order with penalty, final order with no penalty, source agrees, unilateral administrative order without adjudication, combined with another case, or undecided. The outcome may impact both the reporting and penalty determination for each case.

Number of Violations and Laws Violated: For each case, the EPA provides a list of the number of violations under each environmental statute. While most cases focused on violations of the Clean Air Act (CAA), the Clean Water Act (CWA), and/or the Resource Conservation and Recovery Act (RCRA), other environmental statutes were also included in the case information, including the Comprehensive Environmental Response, Compensation, and Liability Act (CERCLA), the Toxic Substances Control Act (TSCA), the Emergency Planning and Community Right-to-Know Act (EPCRA), the Federal Insecticide, Fungicide, and Rodenticide Act (FIFRA), and the Safe Drinking Water Act (SDWA). Cases that included multiple law violations were coded according to the number of laws violated. It is plausible that penalty assessment and media coverage will vary along with the seriousness of a violation, the number of violations, and the type of law violated.

Federal Penalty Sought and Assessed, Compliance Amount, and SEP Amount: Data were collected on the federal penalty sought and the amount assessed in each case. According to the EPA, the compliance amount is "the combination of the injunctive relief and the physical or nonphysical costs of returning to compliance. Injunctive relief represents the actions a regulated entity is ordered to undertake to achieve and maintain compliance, such as installing a new pollution control device to reduce air pollution, or preventing emissions of a pollutant in the first place" (EPA, 2004d). The compliance amount also includes the costs associated with civil court actions. In addition, data on the amount each facility paid into the Supplemental Environmental Project (SEP) was also collected. SEP was enacted by the EPA in order to give the defendant the opportunity to reduce the penalty assessed for a violation. The defendant/respondent agrees to undertake a particular action as stipulated in the order or decree resolving the enforcement action. A SEP is done voluntarily and is negotiated to reduce penalties.

CAA, CWA, and RCRA Information: Data were collected concerning the Clean Air Act, the Clean Water Act, and the Resource Conservation and Recovery Act for major permits, inspections, enforcement actions, penalty assessed, state inspections, current significant non-compliance, and number of quarters of non-compliance over the past two years.

- Permits: Each facility is coded as having a major CAA, CWA, and/or RCRA permit, a minor permit, or no permit.
- Inspections: The number of EPA inspections that have occurred at the facility, under the corresponding statute, within the last two years.
- Enforcement Actions: The number of enforcement actions that have occurred at the facility, under the corresponding statute, within the last two years.
- Penalty Amount Assessed: The amount of penalty assessments that have occurred at the facility, under the corresponding statute, within the last two years.
- State Inspections: The number of state inspections that have occurred at the facility, under the corresponding statute, within the last two years.
- Significant Non-Compliance Violations : Indicates whether or not the facility is in significant non-compliance violation of the corresponding statute within the last two years.
- Quarters of Non-compliance: The number of quarters (out of 8) the facility has been in non-compliance for each statute.
- Demographic Information for Each Facility: For each facility, information was collected for the following demographics; percent minority, percent African-American, percent Hispanic, and percent below poverty within a three mile and five mile radius of the violating facility, and for the county and state where the violation occurred.

LEXISNEXIS News Information

In order to examine news coverage of petroleum refining industry violations, newspaper articles published on cases listed in the EPA's ECHO data between 1997 and 2003 were collected from the LexisNexis database. LexisNexis contains articles from over 25 widely circulated newspapers (see Appendix A). A guided news search was conducted utilizing a wide range of search terms in order to reliably identify news articles covering petroleum industry violations during the search time frame. General search terms included the following: EPA, oil, petroleum, violations, and fines. In addition to general searches, each of the companies included in the ECHO databases were searched for by name in the LexisNexis database. Articles not pertaining directly to the petroleum industry violations under examination were collected in order to provide a more detailed picture of media coverage of the petroleum industry.

Each article was examined for the following information: article location (i.e. front page, business section, etc.); article type (news, editorial, etc.); word count, headline keywords, companies named, and article themes. Articles pertaining directly to cases included in the present study were content analyzed in order to determine which factor led to greater news media coverage of petroleum refining industry violations.

Methods of Analysis

Descriptive statistics were tabulated for research questions one and two in order to describe the nature and distribution of environmental crimes as indicated by federal petroleum violations and the nature and distribution of mainstream news media reporting of the federal petroleum violations.

Research question three involves a content analysis of the news articles that reported on the federal petroleum refining violations included in the present study. The purpose of the content analysis is to describe the factors that led to greater coverage of the violations and to describe the latent content of the news reporting. Content analysis generally involves examining the manifest and latent content of the data. Manifest content refers to the obvious surface content of the data while the latent content refers to the meaning underlying what is stated. Both manifest and latent content analysis were utilized in the present study.

Research question four involved the use of multiple regression. Multiple regression is used to account for (predict) the variance in the dependent variable, based on linear combinations of the independent variables. In other words, multiple regression is utilized to estimate the proportion of the variance in the dependent variable, and determine whether selected independent variables make a significant contribution towards explaining that variance while holding constant competing explanations (i.e., represented by other independent variables). The R^2 can be used to judge the validity of the independent variables as a set of estimators. The variable estimates (b coefficients and constant) are used to construct a prediction equation, and estimate effect sizes.

Multiple regression is based on several underlying assumptions (Pedhazur, 1997): 1) normal distributions, 2) linearity of relationships, and 3) homoscedasticity. Regression assumes that variables have normal distributions and that the relationship between the independent variable and the dependent variable is linear in nature. In order to check for normal distribution and nonlinearity, histograms and scatterplots were examined. The data presented a non-normal distribution and a non-linear pattern. In order to obtain a more normal distribution and provide a better linear fit, logistic transformations of the dependent variables were conducted. Logistic transformation allows for a more normal distribution and linearizes the fit as much as possible (Pedhazur, 1997). The main drawback of log transformations concerns complicating interpretation of the results. Homoscedasticity means that the variance in errors is the same across all levels of the independent variable. In the present study, homoscedasticity was checked through a visual examination of a plot of the standardized residuals by the regression standardized predicted value.

Several regression models were estimated. The dependent variable was the logged penalty difference. Previous research has concentrated on predicting the fine levied by the EPA against oil refineries that violate environmental statutes (Lynch, Stretesky and Burns, 2004a, 2004b). These studies indicate that community race and class characteristics have a significant effect on total EPA penalty assessed. The present study investigates this relationship further by examining the impact of community race and class characteristics on penalty departure. Penalty departure is the difference between the EPA recommended penalty and the final assessed penalty. The distribution of this variable was nonlinear and non-normal. A log transformation approximated a more normal, linear variable.

The independent variables used to predict penalty departure included both legal and extra-legal factors. Legal factors included: voluntary disclosure, number of violations, type of violation (CAA, CWA, RCRA), major and minor violations and permits, number of violations, and Supplemental Environmental Project contributions.

Extra-legal factors consisted of community race, class and ethnic concentration measures (percent minority, percent African-American, percent Hispanic, and percent below poverty) representing the characteristic of people living within three-mile and five mile radii surrounding facilities. Tests for mean racial, ethnic and class variation that measured the difference between county and/or state racial, ethnic and class composition and local area (3 and 5 mile) racial, ethnic and class composition were also tested. These tests were used to assess whether racial, ethnic or class composition per se, or variation in racial, ethnic and class composition relative to larger aggregations (counties and states) might better account for penalty departure.

Study Limitations

Before proceeding to the analysis and results, it is useful to address the limitations of the present research. One of the greatest concerns for researchers studying crime in its various forms is the likelihood that not all crimes are reported, meaning that any official measure of crime contains some measurement bias. For example, researchers routinely conduct crime analyses and make predictions based on the Uniform Crime Reports. While the UCR may provide some of the most reliable statistics on crime in comparison to other surveys (a debatable suggestion), there are still a wide range and number of crimes excluded from the survey. The UCR only reports on *crimes known to police*. The data collected in the present study only report crimes committed by the petroleum refining industry *known to the EPA*. It is widely noted that official reports underestimate the actual amount of crime (Sherman, 1998; MacDonald, 2002); consequently, the nature and distribution of environmental crime as depicted by petroleum refining industry violations may be biased. This problem may be compounded by the fact that EPA enforcement and compliance efforts are heavily influenced by the political climate and budgetary commitments. Consequently, in some years increased enforcement initiatives may be the direct result of political pressure while in other years, budget cuts and other concerns may misdirect environmental concerns.

While secondary data analysis presents a wide range of advantages for social science research inquiry (Bachman and Schutt, 2001), there are also a number of disadvantages which have an impact on the data and analyses in the present study. For a number of cases, the EPA did not report data for a range of variables. In some cases, the EPA case was still open and consequently data was missing for good cause. In other cases, the data was listed as unavailable despite numerous attempts to retrieve the data. Missing data does have an impact on the study results and even more so due to the limited range of cases included in the present study. In the future, researchers engaging in similar research should contact the EPA in order to request missing data. Furthermore, the range of cases can be increased in order to analyze a greater number of cases which lessens the impact of missing data.

Another serious limitation in the present study has to do with the very small number of news articles collected with reference to petroleum refining industry violations. In the future, studies should widen their search parameters in order to increase the potential for wider news coverage.

The present study did not take into consideration other factors that influence penalty assessment decisions. For example, the EPA or the judge (depending on whether the case is administrative or judicial) may base their decisions on personal biases that cannot be readily or easily observed and therefore there is no method by which to control for these other factors.

Finally, the relationships, if any, discovered through the use of regression models cannot demonstrate causality. First, the regression may be inefficient predictors of penalty departure, and important independent variables may have been omitted from consideration. Second, the "causal" relationships measured here cannot be directly observed, but are inferred from the direction and strength of the statistical relationship. For example, if penalty departures are influenced by community class factors, this implies that the EPA has somehow considered community factors in reaching a penalty decision, There is, however, no overt evidence of this influence that can be garnered from the present study of aggregate trends.

CHAPTER 6

Results

INTRODUCTION

The following chapter presents the results for the four research questions discussed in the previous chapter. Overall, the present study found that petroleum refining industry is responsible for a great deal of environmental crimes; media coverage of petroleum refining violations is virtually non-existent; certain factors contribute to the likelihood of news coverage; and that penalty amounts are disproportionately distributed by racial characteristics.

Research Question #1

What is the nature and distribution of environmental crime as indicated by federal petroleum refining violations?

Company Information

The Environmental Protection Agency ECHO database returned one hundred and sixty-two cases. Seventy-eight separate companies were involved in the 162 cases. Of these seventy-eight companies, sixteen companies (20.5% of all companies) were involved in three or more EPA cases (representing a total of 81 cases or 50% of all cases) from 2001-2002 (see Table 4). Twelve companies (15.4% of all companies) were involved in two EPA cases (representing a total of 24 cases or 14.8% of all cases) from 2001-2002 (see Table 5). The remaining fifty companies were involved in one EPA case from 2001-2002.

Table 4: Companies with Three or More EPA Cases 2001-2002 (N=16)

Company Name	Number of Cases
Koch Industries	10
Chevron	9
Shell Oil	8
Motiva Enterprises	7
BP Amoco	6
Marathon Ashland Clark Refining and Marketing	5
Conoco, Inc Cross Oil Refining and Marketing Crown Central Petroleum Sunoco, Inc	4
E.I. DuPont Mobil Oil PRC Patterson Sun Company Inc Tosco Refining Company	3

Table 5: Companies with Two EPA Cases 2001-2002 (N=12)

Company Name
Berry Petroleum
Cyril Petrochemical
Double Eagle Refinery Company
Fina Oil and Chemical
Texaco
Montana Refining Company
Murphy Oil USA
Navajo Refining Company
Phillips Petroleum
Quantum Realty Company
Reichhold Chemicals Inc.
Ultramar Diamond Shamrock Corporation

Violating facilities were located in thirty states across the country with the most cases occurring in Texas (42) followed by Oklahoma (14) and California, Louisiana, and Pennsylvania (each with 11 cases). Table 6 presents the location of the violating facilities by state and percentage that this number represents in the total number of cases.

Table 6: Location of Violating Facility by State and Percentage of Total Cases*(N=162)

State	Number of Cases per state	Percent of Total per state
Texas	42	25.9
Oklahoma	14	8.6
California, Louisiana, Pennsylvania	11	6.8
Illinois	9	5.6
Delaware	8	4.9
Arkansas	7	4.3
Puerto Rico	6	3.7
Minnesota	4	2.5
Michigan, North Dakota, New Jersey, New Mexico, Utah, Virginia	3	1.9

* The following states had two or fewer violating facilities: Arizona, Colorado, Florida, Hawaii, Kansas, Massachusetts, Maryland, Maine, Montana, New York, Ohio, Washington, Wisconsin, and West Virginia.

Overall, federal petroleum refining violation data indicates that a large number of petroleum refining companies are in violation of federal violation statutes. Furthermore, half of the cases (81) involved companies with more than one violation committed from 2001-2002. The data also indicate that violations occur in a majority of states which operate petroleum refining facilities. A more detailed discussion of these results will be presented in Chapter Seven.

Case Type

Cases were coded as either administrative or judicial (civil). Most cases were resolved by the EPA without court intervention (127 cases or 78.4%). The remaining 35 cases (21.6%) involved judicial intervention.

Voluntary Disclosure

For each case, the EPA indicates whether or not the case was the result of a facility self-disclosure. Only 21 cases (13%) involved self-disclosure.

Multi-media

The EPA records whether or not the facility was in violation of more than one environmental statute. Most cases involved one violation (142 cases or 87.7%) while 20 cases (12.3%) involved violations of two or more environmental statutes.

Case Status

Although a large proportion of cases (115 or 71%) were settled or closed from 2001-2002, a number of cases (47 or 29%) were initiated during this time period and remained open or undecided.

Case Outcome

Case outcomes were divided into six categories (see Table 7). Most cases (102 or 63%) received a final order with penalty with the remaining 60 cases falling under one of the five additional categories.

Table 7: Case Outcomes with Frequencies and Percentages (N=162)

Case Outcome	Number	Percent
Final Order with Penalty	102	63.0
Final Order with No Penalty	11	6.8
Source Agrees	9	5.6
Unilateral Administrative Order with No Adjudication	15	9.3
Combined with Another Case	10	6.2
Undecided	15	9.3

Number of Violations

For each case, the EPA provides a list of specific violations committed by the facility for each environmental statute. While the majority of facilities (113 or 69.8%) violated just one environmental statute, the remaining forty-nine facilities were in violation of more than one environmental statute (see Table 8).

Table 8: Number of Violations per Facility with Frequencies and Percentages (N=162)

Number of Violations	Frequency	Percentage of Total
1	113	69.8
2	25	15.4
3	10	6.2
4	7	4.3
5	1	.6
6	5	3.1
9	1	.6

Laws Violated

The majority of cases involved violations of the Clean Air Act (CAA; N = 74; 45.7%), the Clean Water Act (CWA; N = 39; 24.1%) and/or the Resource Conservation and Recovery Act (RCRA; N = 22; 13.6%). Table 9 provides the frequencies and percentages of the six other environmental statutes included in the present study.

Table 9: Frequency and Percent of Environmental Statute Violations (N=202)

Environmental Statute	Frequency	Percentage of Facilities in Violation
CAA	74	45.7
CWA	39	24.1
RCRA	22	13.6
CERCLA	26	16.0
TSCA	12	7.4
EPCRA	26	16.0
FIFRA	2	1.2
SDWA	1	.6

Federal Penalty Sought and Assessed

Data were collected on the amount of the federal penalty sought and the amount assessed in each case. Of the 162 cases, 57 cases (35.2%) did not list information pertaining to penalty sought. The EPA sought a total of $61,788,724 from 105 facilities with a range from $0 to $9,500,000. The average penalty sought was $588,464 when the highest penalty amounts sought (2 x $9,500,000) were included in the calculations. Excluding the two highest penalties sought, the EPA sought a total of $42,788,724 from 103 facilities, or an average penalty sought of $415,424. The average amount sought by the EPA is skewed by the high amounts assessed to a small number of facilities therefore it is important to examine the penalty amount sought by the EPA in terms of the frequency and percentage by dollar range (see Table 10). In 43.2 percent of the cases (70), the EPA sought less than $100,000 in fines. For a small number of cases (12 or 7.5%) the EPA sought more than a million dollars in fines.

Table 10: Penalty Amount Sought By the EPA by Dollar Range (N=105)

Dollar Range	Number of Cases	Percentage of Cases
$0	6	5.7
$1-$10,000	20	19.0
$10,001-$99,999	44	41.9
$100,000-$500,000	18	17.1
$500,001-$1,000,000	5	4.8
$1,000,001-$5,000,000	9	8.6
$5,000,001-$10,000,000	3	2.9

In terms of the federal penalty assessed, 55 cases (34%) did not list information pertaining to penalty assessed. The EPA assessed a total of $49,942,407 from 107 facilities with a range of $0 to $9,500,000. The average penalty assessed was $466,532 when the highest penalty amounts assessed ($9,500,000 and $6,000,000) were included in the calculations. Excluding the two highest penalties assessed, the average penalty assessed was $328,023. Table 11 presents the frequencies and percentages of penalty amount assessed by the EPA by the dollar range. In 48.1 percent of the cases (79), the EPA assessed less than

$100,000 in fines. For a small number of cases (11 or 6.8%) the EPA assessed more than a million dollars in fines.

Table 11: Penalty Amount Assessed By the EPA by Dollar Range (N=107)

Dollar Range	Number of Cases	Percentage of Total Cases
$0	11	10.3
$1-$10,000	36	33.6
$10,001-$99,999	31	29.0
$100,000-$500,000	14	13.1
$500,001-$1,000,000	4	3.7
$1,000,001-$5,000,000	9	8.4
$5,000,001-$10,000,000	2	2.0

Compliance Amount

In 34 of the 162 cases, the EPA assessed compliance costs against the violating facility. Compliance costs include injunctive relief costs and costs associated with returning the violating facility to compliance with EPA statutes. Information pertaining to compliance costs was missing for 50 cases (30.9%) due to case status (open/undecided). Seventy-eight facilities (48.1%) were not assessed any compliance costs. Compliance costs for the remaining 34 facilities (21%) ranged from $5 to $550,000,000. Total compliance costs assessed by the EPA equaled $1,506,698,706. In eighteen of the thirty four cases (52.9%), the EPA assessed compliance costs of $1,000,000 or less (in twelve cases (35.3%), the EPA assessed compliance costs of $5,000 or less). In the remaining sixteen cases (47.1%), the EPA assessed compliance costs of greater than $1,000,000. Of these sixteen cases, 10 cases (29.4%) involved compliance costs between $9,500,000 and $22,000,000 while the highest five compliance cost cases (14.7) were assessed costs ranging from $80,000,000 to $550,000,000. Due to the vast difference in the range of compliance costs, the average compliance cost is misleading ($44,314,668) due to the extremely high amounts assessed to five of the violating facilities. These five facilities alone comprise $1,397,000,000 of the total compliance costs of $1,506,698,000 or 93.7% of the total.

SEP Amount

The Supplemental Environmental Project was enacted by the EPA in order to give the violating facility the opportunity to reduce the penalty assessed for a violation. The violating facility agrees to undertake a particular action as stipulated in the order or decree resolving the enforcement actions. Due to open or undecided cases, SEP amount data was missing for 49 cases (30.2%). No SEP amount was negotiated for 87 cases (53.7%). Twenty-six cases (16.1%) involved negotiation and assessment of an SEP amount. SEP amounts ranged from $1,000 to $7,500,000. The total SEP amount assessed equaled $26,854,509. Of the twenty-six cases assessed SEP amounts, eleven cases (42.3%) were assessed less than $31,000. Eight cases (30.8%) were assessed more than $31,000 but less than $1,000,000. Seven cases (26.9%) were assessed more than $1,000,000 in SEP costs and of those seven cases, three cases (11.5%) were assessed SEP costs in excess of $5,500,000.

CAA, CWA, and RCRA Information

Data were collected pertaining to facility permits, EPA inspections enforcement actions, penalties assessed, state inspections, current significant non-compliance, and number of quarters of non-compliance over the past two years (2003-2004) for each facility, for the Clean Air Act, the Clean Water Act, and the Resource Conservation and Recovery Act. With respect to major permits, number of EPA inspections, number of EPA enforcement actions, and EPA penalty amounts, data were missing for 33 cases (20.4%). With respect to number of state inspections, current significant non-compliance, and number of quarters non-compliance, data were missing for 34 cases (21%). Missing data were the result of undecided cases or a delay in EPA data entry.

Table 12 presents the number and frequencies of CAA, CWA, and RCRA major permit holders in 2003-2004. The majority of the companies were major RCRA permit holders (97.7%) while approximately three-quarters (74.4%) were major CAA permit holders. A little over half (55.0%) of the companies were major CWA permit holders. Sixty-five cases (50.4%) involved companies with all three major permits. Twenty-nine cases (22.5%) involved companies with major CAA and RCRA permits. Ten cases (7.8%) involved companies with major CWA and RCRA permits. There were no cases involving companies with just CAA and CWA major permits.

Table 12: CAA, CWA, and RCRA Major Permits 2003-2004 (N=129)

Major Permit	Number of Cases	Percentage of Cases
CAA, CWA, RCRA	65	50.4%
CAA and RCRA	29	22.5%
CWA and RCRA	10	7.8%
CAA only	2	1.6%
RCRA only	23	17.8%

EPA Inspections

Table 13 presents the number and frequencies of inspections conducted by the EPA from 2003-2004. A large number of companies were not inspected for CAA violations (54.3%), CWA violations (49.6%), or RCRA violations (48.4%). When combining the percentage of inspections for the CAA, CWA, and RCRA, approximately one-third of the companies were inspected by the EPA for CAA, CWA, and RCRA violations from 2003-2004. Eight companies (6.3%) were inspected more than three times for CAA violations; twenty-five companies (19.4%) were inspected more than three times for CWA violations; and thirty companies (23.2%) were inspected more than three times by the EPA for RCRA violations.

Table 13: CAA, CWA, and RCRA Number of EPA Inspections 2003-2004 (N=129)

Number of Inspections	CAA	CWA	RCRA
0	70 (54.3%)	64 (49.6%)	63 (48.4%)
1-2	51 (39.5%)	40 (31.0%)	36 (27.9%)
3-4	6 (4.7%)	15 (11.6%)	15 (11.6%)
5 or more	2 (1.6%)	10 (7.8%)	15 (11.6%)

Enforcement Actions

Table 14 presents the number and frequencies of enforcement actions initiated by the EPA for CAA, CWA, and RCRA violations from 2003-2004. For the CAA, most companies had no enforcement actions (75.2%), although 18 companies (14.0%) had one enforcement action and 14 companies had 2 or more enforcement actions (10.5%). For the CWA, the vast majority of companies had no enforcement actions (96.1%) while 5 companies (3.9%) had one or more enforcement actions. For RCRA, the majority of companies had no enforcement actions (88.3%) although nine companies (7.0%) had one enforcement action and six companies (4.7%) had two enforcement actions.

Table 14: CAA, CWA, and RCRA Number of EPA Enforcement Actions 2003-2004 (N=129)

Number of Enforcement Actions	CAA	CWA	RCRA
0	97 (75.2%)	124 (96.1%)	114 (88.3%)
1	18 (14.0%)	2 (1.6%)	9 (7.0%)
2	5 (3.8%)	2 (1.6%)	6 (4.7%)
3 or more	9 (7.0%)	1 (.7%)	0 (0.0%)

Comparisons of the inspection and enforcement data reveal that 54.2 percent of CAA inspections resulted in an enforcement action, 7.7 percent of CWA inspections resulted in an enforcement action, and 22.7 percent of RCRA inspections resulted in an enforcement action.

Penalty Amounts

Table 15 presents the number and frequencies of penalty amounts assessed by the EPA for CAA, CWA, and RCRA violations from 2003-2004. For the CAA, most companies (82.2%) had no penalty assessments, four companies (3.1%) were assessed less than $10,000 in fines, five companies (3.8%) were assessed fines ranging from $10,001-$100,000, four companies (3.1%) were assessed fines ranging from $100,001-$1,000,000, and one company (.7%) received a fine in excess of $1,000,000. For the CWA, all but two companies (98.4%) received no penalty assessments. One company (.7%) was assessed a penalty of less than $10,000 while the other company received a fine in excess of $1,000,000. For RCRA, most companies (93%) received no fines from the EPA while four companies (3.1%) were assessed less than $10,000 in fines and three companies (2.3%) were ordered to pay fines ranging from $10,000 to $1,000,000.

Table 15: CAA, CWA, and RCRA EPA Penalty Amounts Assessed 2003-2004 (N=129)

Penalty Amount	CAA*	CWA*	RCRA*
$0	106 (82.2%)	127 (98.4%)	120 (93.0%)
$1-$10,000	4 (3.1%)	1 (.7%)	4 (3.1%)
$10,001-$100,000	5 (3.8%)	0 (0.0%)	2 (1.6%)
$100,001-$1,000,000	4 (3.1%)	0 (0.0%)	1 (.7%)
$1,000,001 or more	1 (.7%)	1 (.7%)	0 (0.0%)

* Range of penalty amounts for CAA: $0-$4,395,407; CWA: $0-$4,500,000; RCRA: $0-$205,866

Table 16 presents the number and frequency of inspections conducted at the state level for the CAA, CWA, and RCRA from 2003-2004. State inspections were conducted more frequently than federal inspections. For the CAA, forty companies (31.3%) had no inspections, 31 companies (24.2%) had 1-3 inspections, eighteen companies (14.1%) had from 4-6 inspections, and 31 companies (25%) had more than 7 inspections. For the CWA, over half of the companies (53.1%) were never inspected by the state from 2003-2004. Forty-eight companies (37.5%) were inspected from 1-3 times and 12 companies (9.4%) were inspected over 4 times by the state. For RCRA, almost half of the companies (49.2%) were not inspected by the state while forty-four companies (34.4%) had 1-3 inspections. Twenty-one cases (16.4%) were inspected by the state more than four times from 2003-2004.

Table 16: CAA, CWA, and RCRA Number of State Inspections 2003-2004 (N=128)

Number of Inspections	CAA	CWA	RCRA
0	40 (31.3%)	68 (53.1%)	63 (49.2%)
1-3	31 (24.2%)	48 (37.5%)	44 (34.4%)
4-6	18 (14.1%)	1 (.8%)	15 (11.7%)
7-9	9 (7.0%)	1 (.8%)	6 (4.7%)
10 or more	23 (18.0%)	10 (7.8%)	0 (0.0%)

When examining all inspections (by both the EPA and the state), the data reveals the following: in thirty cases (23.3%) no inspections were conducted by the EPA for CAA, CWA, or RCRA violations from 2003-2004; in thirty-nine cases (24.1%) no inspections were conducted by the EPA or the state for CAA violations from 2003-2004; in sixty-three cases (48.8%) no inspections were conducted by the EPA or the state for CWA violations from 2003-2004; and in sixty-two cases (48.1%) no inspections were conducted by the EPA or the state for RCRA violations from 2003-2004.

Significant Noncompliance

Table 17 presents the number and frequency of companies determined by the EPA to be in significant non-compliance with the CAA, CWA, and RCRA. Sixty companies (46.9%) were in significant non-compliance with the CAA, four companies (3.1%) were in significant non-compliance with the CWA, and eight companies (6.3%) were in significant non-compliance with RCRA.

Table 17: CAA, CWA, and RCRA Significant Non-Compliance 2003-2004 (N=128)

Significant Non-Compliance	CAA	CWA	RCRA
YES	60 (46.9%)	4 (3.1%)	8 (6.3%)

Table 18 presents the number and frequency of the quarters of non-compliance (out of 8) for the CAA, CWA, and RCRA from 2003-2004. Non-compliance can result from three conditions: (1) the company is found to be in current noncompliance with statutes; (2) the company has failed to remedy a past non-compliance finding; (3) the company has failed to file a compliance statement with the EPA. Sixty-two companies (48.4%) were in non-compliance for the CAA for 7 or 8 quarters while fifty-two companies (40.6%) had zero quarters of non-compliance. Half of the companies (50.0%) had zero quarters of non-compliance for the CWA while fifteen companies (11.7%) were in non-compliance for 7-8 quarters. Seventy-five companies (58.6%) had zero quarters in non-compliance with RCRA while thirty-five companies (27.3%) were in non-compliance for 7-8 quarters.

Table 18: CAA, CWA, and RCRA Number of Quarters of Non-Compliance 2003-2004 (N=128)

Quarters of Non-Compliance	CAA	CWA	RCRA
0	52 (40.6%)	64 (50.0%)	75 (58.6%)
1-2	2 (1.6%)	18 (14.1%)	12 (9.4%)
3-4	7 (5.5%)	20 (15.6%)	4 (3.1%)
5-6	5 (3.9%)	11 (8.6%)	2 (1.6%)
7-8	62 (48.4%)	15 (11.7%)	35 (27.3%)

Summary of Results for Research Question #1

Results of the descriptive statistics tabulated for research question #1 on the nature and distribution of environmental crime as indicated by federal environmental violations committed by the petroleum refining industry indicate the following:

- Violations of environmental statutes are frequent and widespread.

- Thirty-six percent of companies were involved in more than one EPA case from 2001-2002.

- The majority of states (thirty) hosting petroleum refining operations had a least one refinery in violation of environmental statutes.

- One out of every five cases involved judicial intervention.

- Only a small number of cases (13%) involved a facility self-disclosure of violations.

- The Clean Air Act was the most frequently violated statute.

- Over half (50.4%) of cases involved companies with major permits for the CAA, CWA, and RCRA.

- Over twenty-three percent of cases involved companies with no inspections by the EPA or the state for violations of the CAA, CWA, or RCRA from 2003-2004.

- Forty-seven percent of cases involved companies in significant non-compliance of the CAA from 2003-2004.

- Forty-eight percent of cases involved companies in non-compliance with the CAA for 7 or quarters of 2003-2004.

Research Question #2

What is the nature and distribution of mainstream news media reporting of federal petroleum refining violations?

News articles from the LexisNexis database were collected from 1997 to 2003 corresponding with the earliest EPA initiated case in 1997 and a year after the last case was initiated in 2002. A guided news search was conducted in order to obtain the expanse of news articles covering petroleum refining violations during the search time frame. Each article was examined for the following information; article location, article type, word count, and case match. In addition, a content analysis was conducted in order to determine which factors lead to greater news media coverage of petroleum refining industry violations.

Seventy-four articles were collected with reference to petroleum refining industry violations. Of these seventy-four articles, seventeen articles (23%) corresponded directly with cases included in the EPA ECHO database. The remaining fifty-seven articles (77%) reported on the petroleum refining industry but were not directly related to any of the cases included in the ECHO database.

All News Articles on the Petroleum Refining Industry

Seventy-four news articles were collected with respect to the petroleum refining industry. Table 19 presents the year and number of articles collected during that year. The majority of articles appeared in 2000 (22 or 29.7%) or 2001 (25 or 33.8%), representing 63.5% of the total number of articles.

Table 19: Number of News Articles on the Petroleum Refining Industry by Year (N=74)

YEAR	Number of Articles
1997	5
1998	2
1999	0
2000	22
2001	25
2002	9
2003	11

Articles appeared in twenty different news sources. Table 20 presents the news sources and the number of articles presented by each source. The Houston Chronicle produced the most news articles (15 or 20.3%) followed by the Times Picayune (News Orleans) with 12 articles (16.2%) and the Star Tribune (Minneapolis) and the New York Times with 7 articles each (9.5%). Together, articles from these four news sources (41) represent over half (55.5%) of the total number of articles.

Table 20: News Sources and Number of Articles by Source (N=74)

News Source	Number of Articles	News Source	Number of Articles
Atlanta Journal Constitution	2	San Antonio Express	1
Chicago Sun-Times	5	San Diego Union Tribune	1
Columbus Dispatch	1	San Francisco Chronicle	5
Daily News (New York)	2	Seattle Times	1
Denver Post	1	St. Louis Post	2
Houston Chronicle	15	Star Tribune (Minneapolis)	7
Milwaukee Journal Sentinel	3	Tampa Tribune	2

Table 20 continued: News Sources and Number of Articles by Source (N=74)

News Source	Number of Articles	News Source	Number of Articles
New York Times	7	Times Picayune (New Orleans)	12
Pittsburgh Post Gazette	1	USA Today	1
Rocky Mountain News (Denver)	2	Washington Post	3

Each article was examined for article location. News articles appeared in one of four locations: 1) News (including Section A and National News), 2) Local News (including Section B, Metro, and Suburban), 3) Business (including Money), and 4) Other (including editorials or Science). Table 21 presents the frequencies and percentages of news articles by the article location. For each article location, information was also gathered according to location of the article within that specific sub-section of the paper. Table 22 presents the frequencies and percentages of article locations within each of the four major categories. Over forty-four percent of the articles (33) appeared on the front page of the paper sub-section while over forty percent of the articles appeared on page four or higher (30).

Table 21: Frequencies and Percentages of News Articles by Article Location (N=74)

Article Location	Number and Percent of Total
News (Section A/National)	29 (39.2%)
Local (Section B/Metro)	21 (28.4%)
Business (Money)	21 (28.4%)
Other (Editorial/Science)	3 (4.5%)

Table 22: Frequencies of News Articles by Sub-Section of Article Location (N=74)

Location	Front page	Page 2 or 3	Page 4 or higher
News	8	2	19
Local	11	5	5
Business	14	4	3
Editorial	0	0	3
Total	33 (44.6%)	11 (14.9%)	30 (40.5%)

The majority of the articles were considered strictly "news" pieces (67 or 90.5%) while a small number of articles were considered either "news briefs" (5 or 6.8%) or "editorials" (2 or 2.7%).

Data were collected on the total word count for each article. The lengthiest article contained 5,729 words while the shortest article contained just 79 words. The average word count including the lengthiest article was 630 words per article but excluding the lengthiest article the average word count drops to 568 words per article. Fifty-two articles (70.3%) contained less than 700 words while the remaining twenty-two articles contained 700 words or more (29.7%). Table 23 presents the frequencies and percentages of news articles by word count category. News papers have varying guidelines in terms of column length and word count. On average, feature or front section news stories will contain just over 1,000 words suggesting that the articles concerning environmental crime are about half as long (568 words per article on average).

Table 23: Frequencies and Percentages of News Articles by Word Count (N=74)

Number of Words	Number of Articles	Percentage of Articles
Less than 100	4	5.4
101-300	15	20.3
301-500	15	20.3
501-700	18	24.3
701-900	11	14.9
901-1,100	6	8.1
1,100 or more	5	6.8

Case Specific News Articles on the Petroleum Refining Industry

The news data presented in the preceding paragraphs and in Tables 19 through 23 are reflective of all news articles collected on the petroleum refining industry from 1997 to 2003. A small number of articles (17 or 23%) pertained directly to one or more of the cases presented in the ECHO databases. Table 24 presents information pertaining to the company named in the article, the ECHO case number, and the number of articles addressing that specific case.

Although only seventeen articles were directly related to the cases analyzed in the present study, a number of articles mentioned more than one company. For example, BP and Koch were reported on in four articles with Koch also mentioned in one additional article. Both Marathon Ashland and Motiva were reported on in three articles for each company. Conoco (3 articles), Clark Refining and Marketing (2 articles), and Murphy Oil (1 article) were the only singular cases to be reported on in the news articles.

Table 24: Company Name, ECHO Case Number, and Number of News Articles (N=17)

Company	ECHO Case Number	News Articles
Koch Petroleum Group	31	5*
Koch Petroleum Group	55	5*
Koch Petroleum Group	56	5*
BP Exploration and Oil Company	32	4*
BP Exploration and Oil Company	33	4*
Marathon Ashland Petroleum	61	3**
Marathon Ashland Petroleum	29	3**
Marathon Oil Company, Inc.	28	3**
Motiva Enterprises, LLC	62	3***
Motiva Enterprises, LLC	63	3***
Motiva Enterprises, LLC	64	3***
Conoco, Inc.	97	3
Clark Refining and Marketing Inc.	22	2
Murphy Oil USA, Inc.	24	1

* All five cases concerning BP and Koch were represented in four of the same articles (Koch had one additional article).

** All three cases concerning Marathon Ashland were represented in the three corresponding news articles.

*** All three cases concerning Motiva were represented in the three corresponding news articles.

ECHO Case Information and Corresponding Article Information

Data are presented on the specific ECHO cases that appeared in news articles relating to the violation. In addition, information from each article is presented below. A more detailed discussion of the latent content of the articles and case comparisons will be examined in the following Discussion Chapter.

Clark Refining and Marketing Inc. (ECHO Case #22) (Two Articles)

According to the EPA, Clark Refining and Marketing Inc. violated the CAA, CWA, RCRA, CERCLA, and EPCRA, at their refinery located in Blue Island, Illinois. The EPA filed a civil case against the company on September 9, 1998 and settled the case on June 12, 2002. The company did not voluntarily disclose the violations. The penalty amount sought by the EPA was not disclosed but the EPA assessed the company $3,125,000 for the violations in addition to $1,450,000 in compliance costs.

Two news articles (from the St. Louis Dispatch and the Washington Post) appeared on the date following the filing of the case by the EPA (9/10/1998). The first news article appeared in the St. Louis Dispatch Metro section (word count = 717) and reported on a press conference attended by then Attorney General Janet Reno who announced that the Justice Department would be pursuing civil action against several oil companies for EPA violations, including Clark Refining and Marketing. The case was mentioned as part of the Mississippi River Initiative, a federal effort to protect the Mississippi River. For the most part, the article focused on violations committed by Shell Oil Company (not included in the ECHO database) but did mention that Clark was responsible for violating a number of federal statutes including the Clean Air Act and the Clean Water Act. Reactions from Clark officials were presented in the article. No penalty amounts were included in the article.

The news article from the Washington Post, Section A, page 3, (word count = 1,001) was similar in content to the news article presented in the St. Louis Dispatch. Again, quotes from then Attorney General Janet Reno were reported and an emphasis was placed on Shell Oil Company. Information pertaining to pollution of the Mississippi River was highlighted. Enforcement actions against Clark Refining and Marketing were mentioned very briefly, in just one sentence of the entire article.

Conoco, Inc. (ECHO Case #97) (Three Articles)

According to the EPA, Conoco, Inc. violated the CAA at their refinery located in Ponca City, Oklahoma and at eight other refineries located across the United States, including a refinery in Commerce City, Colorado. The EPA filed a civil case against the company on December 21, 2001 and settled the case on February 20, 2002. The company did not voluntarily disclose the violation. The EPA sought $1,500,000 in fines but assessed the company $824,000 in fines in addition to compliance costs of $80,000,000 and an SEP amount of $1,800,000.

Three articles appeared on the same day the case was filed by the EPA (12/21/01). Article sources included the Houston Chronicle, the Denver Post, and the Rocky Mountain News (Denver). The news article from the Denver Post Metro section (word count = 536) focused on Conoco's violation of the CAA at the Commerce City, CO refinery. The article stated that Conoco agreed to pay over $22 million on air pollution controls at the Commerce City refinery and also mentioned a $145,000 civil penalty and $2.6 million SEP amount. State health officials, regional EPA officials, and community leaders were quoted in the body of the article. Emphasis was placed on Conoco's quick reaction to comply with regulations once the company became aware of the violations. Although the case against Conoco was judicial in nature, article statements imply that Conoco agreed to work with the EPA in order to avoid lengthy litigation.

The second news article concerning Conoco appeared in the Rocky Mountain News, another Denver, Colorado publication. The article was reported in the Local section of the paper and contained 703 words. The body of the article was very similar to the information reported in the Denver Post. Penalty amounts were reported and local officials were quoted. Again, emphasis was placed on Conoco's quick response to the pollution allegations and subsequent compliance efforts. The article described the benefits to the community in terms of decreased air emissions and support for local environmental projects.

The third and final article addressing Conoco appeared in the Houston Chronicle Business section (word count = 243). The brief article stated that Conoco agreed to spend up to $110 million to reduce emissions at its various refineries located in Louisiana, Oklahoma, Colorado, and Montana. Similar to the two previous articles, the Chronicle article also reported on a $1.5 million civil penalty and a $5 million SEP amount. While the majority of the article addressed the EPA violations, the article also discussed Conoco's intentions to purchase interests in a natural gas project off the Vietnam coast.

Marathon Ashland (ECHO Cases #28, #29, and #61) (Three Articles)

According to the EPA, Marathon Ashland violated the CAA, CWA, RCRA, and TSCA at their refineries located in Texas, Illinois, Louisiana, Kentucky, Ohio, Minnesota, and Michigan. The judicial case was filed by the EPA on May 11th, 2001 and settled on August 28th, 2001. The EPA combined the cases for penalty assessment purposes. No cases were voluntarily disclosed. The EPA sought $3,800,000 in penalties and assessed $3,700,000 in penalties. In addition, Marathon Ashland was assessed $265,000,000 in compliance costs and agreed to pay $6,500,000 in SEP amounts.

Three articles concerning the Marathon Ashland cases appeared the day after the EPA initiated its judicial cases against the company (5/12/01). Articles appeared in the St. Louis Dispatch, the Star Tribune (Minneapolis), and the Houston Chronicle. The article appearing in the St. Louis Dispatch was part of a section entitled "Nation and World Briefs"; consequently the information reported was very brief. The article stated that Marathon Ashland was assessed an estimated $265 million to install pollution control equipment at its refineries. The article listed the location of the violating refineries and stated that Marathon Ashland was responsible for a $3.8 million civil penalty and a $6.5 million SEP amount.

The second article addressing Marathon Ashland was reported in the News section of the Star Tribune (Minneapolis) (word count = 607). The article stated that Marathon Ashland had "agreed" to spend $265 million on pollution control equipment, a $3.8 million civil penalty, and a $6.5 million SEP amount. In addition, the article repeatedly addressed an "odor" problem affecting the community located near the St. Paul Park refinery. The article reported the various ways in which Marathon Ashland is required to reduce pollution at its refineries and stated that several of the pollutants "have been associated with serious respiratory problems and could exacerbate childhood asthma". Although the case was not voluntarily disclosed, a Marathon Ashland official stated that the settlement was voluntary and would avoid litigation. The article lists the other refineries included in the settlement and describes the lengthy violation history of the St. Paul Park Refinery.

The third and final article concerning the Marathon Ashland cases appeared in the Houston Chronicle Business section (word count = 231). Like the first article from the St. Louis Dispatch, the article is considered a "news brief". The article begins by stating that the cases against Marathon Ashland were "settled" on the previous day although according to EPA data, the cases were actually filed on that day and settled several months later. Again, similar to the two previous articles, the Chronicle article reports penalty amounts and refinery locations. The article also addressed how new equipment would "help ease respiratory problems such as childhood asthma". Attorney General John Ashcroft is quoted as stating that the settlement was "a victory for the environment".

Motiva Enterprises (ECHO Cases #62, #63, and #64) (Three Articles)

According to the EPA, Motiva Enterprises violated the CAA, CWA, RCRA, TSCA, and EPCRA at their refineries located in Louisiana and Texas. The judicial case was filed by the EPA on March 21st, 2001. The EPA combined the cases for penalty assessment purposes. No cases were voluntarily disclosed. The EPA sought $4,400,000 in penalties and assessed $4,400,000 in penalties. In addition, Motiva was assessed $400,000,000 in compliance costs and agreed to pay $5,500,000 in SEP amounts.

Three news articles appeared on the day following the initial filing of the case by the EPA (3/22/01). Articles were reported in the Houston Chronicle, the Time Picayune, and the Seattle Times. The Houston Chronicle article appeared in the News section, front page, and contained 733 words. The article immediately reported the high compliance amount ($400 million) assessed against Motiva for violations at nine refineries located in five states. The article also mentioned a penalty amount of $9.5 million and an SEP amount of $5.5 million and contained quotes from the EPA administrator and officials from the Department of Justice. While the article addresses violations committed by Motiva, other violating refineries are also included. Descriptions of the air pollutants released by the offending facilities are reported although there is no discussion of the related health concerns.

The second article addressing Motiva appeared in the Times Picayune National news section (word count = 1,005). Again, like the previous article concerning Motiva, penalty, compliance, and SEP amounts are reported. Quotes from prominent Louisiana environmental officials appear throughout the article. Then Attorney General John Ashcroft stated, "protecting our natural resources. . .is a top priority for the Department of Justice". No mention is made about how human health is affected by pollution. The article addressed a criminal case pending against Motiva concerning alleged faulty record keeping.

The third and final article addressing Motiva appeared in the Seattle Times "News Across the Nation" section (word count = 443). The article summarized the cases against Motiva and included a list of the offending refineries, the penalty/compliance/SEP amounts, and the amount of emissions that would be reduced under the case agreement.

BP and Koch (ECHO Cases #31, #32, #33, #55, and #56)

According to the EPA, BP Amoco violated the CAA and RCRA at their refineries located in CA, UT, LA, WA, ND, OH, IN, OK and VA. The judicial cases were filed by the EPA on January 18th, 2001 and settled on August 29th, 2001. The EPA combined the cases for penalty assessment purposes. The EPA sought $9,500,000 in penalties and assessed $9,500,000 in penalties. In addition, BP was assessed $550,000,000 in compliance costs. According to the EPA, Koch violated the CAA, CWA, RCRA, and EPCRA at their refineries located in Texas and Minnesota. The judicial cases were filed by the EPA on December 22nd, 2000 and settled on April 25th, 2001. The EPA combined the cases for penalty purposes. The EPA sought $4,500,000 in penalties and assessed $4,500,000 in penalties. In addition, Koch was assessed $102,000,000 in compliance costs. These cases are described together due to the fact they were reported on simultaneously in the following news articles.

Four news articles addressed the violations committed by BP and Koch and one addition article addressed violations committed solely by

Koch. Articles appeared in the Houston Chronicle, the Washington Post, the Times Picayune, and the Star Tribune (2). The article reported on in the Houston Chronicle Business section (word count = 538) listed the 12 violating refineries and 10 affected states including local refineries in Texas City and Corpus Christi. Penalty amounts, compliance costs, and SEP amounts are reported and the article emphasized that the almost $600 million compliance costs were the "largest enforcement agreements" related to air pollution in history. Quotes from the EPA administrator were included and the companies were hailed for voluntarily initiating negotiations (although according to the EPA the violations were not self-reported). The article included a description of the agreement stipulations and highlighted BP and Koch's 15 percent total U.S. oil refining capacity. The companies were praised for their "cooperativeness".

The second article concerning BP and Koch appeared in the Washington Post in a "news brief" (word count = 115). The high compliance costs ($600 million) are reported. The twelve violating refineries are lists. And again, BP and Koch's 15% of total oil refining capacity is reported.

The third article concerning BP and Koch appeared in the Times Picayune Money section (word count = 708). The article is very similar to the above mentioned articles in that it reported penalty, compliance, and SEP amount, location of violating facilities, the 15% oil refining capacity, and descriptions of agreement stipulations. The article quotes the EPA administrator and violating company officials.

The fourth article concerning BP and Koch appeared in the Star Tribune News section (word count = 825). The article included penalty amounts, compliance amounts, and quotes from the EPA administrator as well as the impact of emission reductions on the environment. Again, no mention of human health impacts were addressed.

The fifth article addressed violations committed solely by Koch and appeared in the Star Tribune News section (word count = 276). The article reported on the penalty and compliance amounts as well as the reduction in terms of air pollutants.

Summary of Results for Research Question #2

News articles on petroleum refining industry violations are similar in terms of reporting trends and themes. The overall image of federal petroleum refining violations as reported by news articles is misleading. Articles liberally quote government officials and company representatives but rarely, if ever, quote residents or victims of environmental crime. The overabundance of quotes from government sources suggests the problem is receiving a great deal of attention and that enforcement is a priority. Violations are reported in vague and general terms. An emphasis was placed on penalty amounts and compliance costs. Companies are often praised for cooperating with government officials. Environmental and human health concerns are either ignored or downplayed. Articles do not equate environmental violations with criminal behavior.

Research Question #3

Which factors lead to greater news coverage of petroleum refining violations?

There is very little news coverage of federal petroleum refining violations. Very few ECHO cases receive any attention from the mainstream news media. Of seventy-four articles concerning petroleum refining industry violations from 1997 to 2003, only 17 or 23% were directly related to the ECHO cases analyzed in the present study. Overall, lack of news media coverage of federal petroleum refining industry violations suggests that the media considers such law violations as rarely newsworthy. Although newspaper coverage of federal petroleum refining violations is lacking, the news articles that were reported suggest that certain factors lead to the likelihood of greater news article coverage. These major factors include: 1) Initial date ECHO case was filed by the EPA, 2) Location of the violating refinery, 3) Large penalty, compliance, and SEP amounts assessed to the violating company by the EPA, and 4) Refining capacity.

Initial Date of Case Filed by the EPA

Of the seventeen articles addressing specific ECHO cases included in the present study, twelve articles (70.6%) appeared in news sources on the same day the EPA case was filed or in the week immediately following the initial case filing. In five articles (29.4%), the date the case was filed or settled appeared to have no direct correlation with the date the article appeared in the news source. No articles appeared on the day the case was settled by the EPA or on days immediately following the settlement. Table 25 presents the name of the company, the news article source, the case filing date, and the date the article appeared for the twelve cases with article matches for the date the case was filed.

Table 25: Company, News Article Source, Case Filing Date, and Date Article Appeared (N=12)

Company	News Article Source	Date Case Filed	Date of Article
Motiva	Houston Chronicle	3/21/01	3/22/01
Motiva	Times-Picayune	3/21/01	3/22/01
Motiva	Seattle Times	3/21/01	3/22/01
Koch	Star Tribune	12/22/00	12/27/00
Marathon Ashland	Houston Chronicle	5/11/01	5/12/01
Marathon Ashland	Star Tribune	5/11/01	5/12/01
Marathon Ashland	St. Louis Dispatch	5/11/01	5/12/01
Conoco	Houston Chronicle	12/21/01	12/21/01
Conoco	Rocky Mountain News	12/21/01	12/21/01
Conoco	Denver Post	12/21/01	12/21/01
Clark Refining and Marketing	St. Louis Dispatch	9/9/98	9/10/98
Clark Refining and Marketing	Washington Post	9/9/98	9/10/98

Location of the Violating Refinery

News articles addressing federal petroleum industry violations were more likely to appear if a violating refinery was located in the community in which the news source was primarily distributed. Thirteen articles (76.5%) addressing the ECHO cases reported on the cases with specific references to problems caused by local refineries. Only four news articles (23.5%) appeared in news publications with no ties to local violating refineries and two of these four articles appeared in the Washington Post. Table 26 presents the name of the company, news source, and the location of the violating refinery for the thirteen articles with local refineries.

Table 26: Company, News Source, News Source City/State, and Location of Violating Refinery (N=14)

Company	News Source	City/State	Location
Conoco	Denver Post	Denver, CO	Commerce City, CO
Conoco	Rocky Mountain News	Denver, CO	Commerce City, CO
Conoco	Houston Chronicle	Houston, TX	Houston, TX*
Marathon Ashland	Star Tribune	Minneapolis, MN	St. Paul Park, MN
Marathon Ashland	Houston Chronicle	Houston, TX	Texas City, TX
Murphy Oil	Milwaukee Journal Sentinel	Milwaukee, WI	Superior, WI
Motiva	Houston Chronicle	Houston, TX	Port Arthur, TX
Motiva	Seattle Times	Seattle, WA	Anacortes, WA
Motiva	Times Picayune	New Orleans, LA	Norco, LA
BP and Koch	Houston Chronicle	Houston, TX	Texas City/ Corpus Christi, TX
BP and Koch	Times Picayune	New Orleans, LA	Belle Chasse, LA
BP and Koch	Star Tribune	Minneapolis, MN	Rosemont, MN
Koch	Star Tribune	Minneapolis, MN	Rosemont, MN

* Conoco had no violating facilities in Houston but the company headquarters are located in Houston.

Large Penalty Amounts

In sixteen (94%) of the seventeen articles directly related to the ECHO cases, EPA penalty sought, compliance amount, and SEP amount were reported. The cases reported on in news articles were all ranked as the top cases in terms of the penalty sought, the compliance amount, and SEP amount. Tables 27 presents the company, EPA penalty sought, and rank among all 162 ECHO cases in terms of the amount of the penalty sought. Companies reported on in news articles were all ranked in the top ten (aside from Conoco which ranked 11) in terms of penalty amount sought. BP had the highest penalty amount sought ($9,500,000). Overall, cases reported on in news articles represented seven of the highest eleven penalty amounts sought by the EPA.

Table 27: Company, Federal Penalty Sought, and Rank of Penalty Amount (N=162)

Company	Penalty Sought	Rank of Penalty Amount
BP	$9,500,000	1
Murphy Oil	$9,500,000	2
Koch	$4,500,000	4
Motiva	$4,400,000	5
Clark Refining and Marketing	$4,000,000	6
Marathon Ashland	$3,800,000	7
Conoco	$1,500,000	11

Table 28 presents the company, compliance amount, and rank among all 162 cases in terms of the compliance amount. Companies reported on in news articles were all ranked in the top ten (aside from Murphy Oil which ranked 12) in terms of compliance amount. Overall, cases reported on in news articles represented seven of the highest twelve compliance amounts.

Table 28: Company, Compliance Amount, and Rank of Compliance Amount (N=162)

Company	Compliance Amount	Rank of Compliance Amount
BP	$550,000,000	1
Motiva	$400,000,000	2
Marathon Ashland	$265,000,000	3
Koch	$102,000,000	4
Conoco	$80,000,000	5
Clark Refining and Marketing	$22,000,000	6
Murphy Oil	$4,500,000	12

Table 29 presents the company, SEP amount, and rank among all 162 cases in terms of the SEP amount. Five of the seven companies reported on in news articles were ranked in the top ten in terms of SEP amount. Overall, cases reported on in news articles represented five of the highest six compliance amounts.

Table 29: Company, SEP Amount, and Rank of SEP Amount (N=162)

Company	SEP Amount	Rank of SEP Amount
Murphy Oil	$7,500,000	1
Marathon Ashland	$6,500,000	2
Motiva	$5,500,000	3
Conoco	$1,800,000	4
Clark Refining and Marketing	$1,200,000	6
BP	$0	---
Koch	$0	---

Table 30 presents the company, the total amount in terms of penalty assessed, compliance amount and SEP amount, and the rank of amount among all 162 cases in the ECHO data included in the present study. The companies reported on in news articles were *the* highest seven cases in terms of total penalty assessed, compliance amount, and SEP amount. News articles only reported on cases in which the total

EPA costs exceeded $20,000,000. Consequently, cases involving lower penalties, compliance costs, and SEP amounts did not receive any news coverage. It appears that the summation of the penalty amount assessed, compliance amount, and SEP amount is the greatest factor leading to news article coverage of petroleum refinery industry violations.

Table 30: Company, Total Penalty Sought, Compliance Amount, SEP Amount, and Rank of Amount

Company	Total Amount	Rank
BP	$559,500,000	1
Motiva	$409,900,000	2
Marathon Ashland	$275,300,000	3
Koch	$106,500,000	4
Conoco	$83,300,000	5
Clark Refining and Marketing	$27,200,000	6
Murphy Oil	$21,500,000	7

Refining Capacity

The greater the refining capacity of the company, the greater the likelihood of news coverage of petroleum industry violations. In ten (59%) of the seventeen articles, the total refining capacities of the violating company were reported. Taken together, just four companies represent 30 percent of the total refining capacity in the United States (See Table 31). It appears that news coverage of petroleum industry violations is more likely to occur if the violating company is responsible for a high percentage of the total United States refining output.

Table 31: Company, News Source, Refining Capacity, and Number of Violating Refineries

Company	News Source	Refining Capacity (total in U.S.)	Number of Violating Refineries
BP and Koch	Star Tribune Times Picayune Washington Post Houston Chronicle	15%	12

Table 31 continued: Company, News Source, Refining Capacity, and Number of Violating Refineries

Company	News Source	Refining Capacity (total in U.S.)	Number of Violating Refineries
Motiva	Times Picayune Seattle Times Houston Chronicle	10%	9
Marathon Ashland	Houston Chronicle Star Tribune St. Louis Dispatch	5%	7

Summary of Results for Research Question #3

News coverage of federal petroleum refining violations is limited. Only seventeen articles from 1997 through 2003 reported on cases included in the EPA's ECHO database. While coverage is lacking, the results presented in the previous sections suggest that certain factors lead to the likelihood of news coverage. Most articles appeared on the day or days immediately following the filing of the case by the EPA, suggesting the EPA press releases are a prevalent source of environmental violation news for reporters. News coverage was also directly linked to violations committed by local refineries, indicating that environmental crime in the form of federal petroleum refining violations are of local concern rather than national concern. The greatest factor leading to news coverage appears to be penalty and compliance amounts. The seven companies receiving the highest penalty amounts by the EPA were also the only seven companies reported on in news articles. In addition, the higher the refining capacity of the company, the greater the likelihood of reporting. A more detailed discussion of these factors is presented in Chapter Seven.

Research Question #4

Are petroleum refining industry penalty assessment decisions affected by the racial and socioeconomic composition of the communities surrounding the violating facility? Do other factors influence penalty assessment decisions?

Originally, it was the intention of this research to determine the impact of community race and ethnic characteristics on penalties assessed against oil refineries that violated the environmental laws. Preliminary analysis revealed that the assessed fine was largely predicted by the preliminary fine requested by the EPA (the Pearson correlation coefficient between penalty assessment and penalty sought by EPA was .931). Community race characteristics were not significant predictors of either the preliminary or assessed penalty. Research on racial bias in criminal justice processes helps to explain this outcome. Racial biases are more likely to be seen in early processing stages. Late stage decision making, particularly sentencing decisions, tends to involve a biased sample produced by earlier stage decisions (see Lynch and Patterson, 1991, 1996). With respect to EPA decisions, for example, it is plausible that a larger number of cases affecting minority communities are excluded before the penalty assessment stage, while less serious cases affecting white communities are retained. These processing selection biases make race appear to be an unimportant determinant of assessed penalties. To address whether or not this is indeed a valid explanation, stage related case processing data would be needed. Data of this nature is not public record, and would require special access to EPA files to collect.

Given the limitations described above, the analysis was redirected to determine if community race, ethnic and class characteristics impacted the departure from the EPA penalty recommendation. In other words, are community characteristics related to the penalty difference controlling for other legally relevant decision making criteria that might impact penalty assessment.

Several regression models were estimated in order to answer this revised question. The dependent variable for each of the regression models presented below was the logged penalty difference. As stipulated in Chapter Five, the distribution of this variable was nonlinear and non-normal. A log transformation approximated a more normal, linear variable.

Table 32 presents the results of the first regression model predicting the percent of departure in penalty amount. The regression equation included the following independent variables: number of violations (vionum), voluntary disclosure (voldis), and percent minority (African-American and Hispanic) within a three mile radius of the violating facility (minper3).

Table 32: Regression Model 1 Predicting Penalty Difference

Variable	B	SE b	Beta	VIF	T	Sig. T*
minper3	-6.432	.004	-.190	1.002	-1.687	.097
vionum	.448	.105	.478	1.004	4.248	.000*
voldis	19.200	.803	-.013	1.005	-.115	.909

p<.05, R^2 = .269, Adj. R^2 = .231,

S.E. = .79505, Durbin-Watson = 1.637

The R^2 for the first regression model presented in Table 32 was .269. The adjusted R^2 was .231, indicating that the independent variables predicted approximately twenty-three percent of the variation in penalty departures. The results from model 1 indicate that whether or not a company voluntarily discloses a violation has no significant effect on the penalty difference. Likewise, the percent minority within a three-mile radius had no significant effect on the penalty difference. The number of violations committed, however, was a statistically significant predictor of penalty departure (p = .000).

Table 33 presents the results of the second regression model predicting the percent of departure in penalty amount where the percent African-American population within a three-mile radius was entered in place of percent minority population (which include Hispanics within a three-mile radius) to determine if there are specific race effects. The regression equation included the following independent variables: number of violations (vionum), voluntary disclosure (voldis), and percent African-American within a three mile radius of the violating facility (aaper3).

Table 33: Regression Model 2 Predicting Penalty Difference

Variable	B	SE b	Beta	VIF	t	Sig. T*
aaper3	-1.295	.005	-.312	1.027	-2.865	.006*
vionum	.407	.102	.434	1.028	3.976	.000*
voldis	-.149	.771	-.021	1.006	-.193	.847

$p<.05$, $R^2 = .328$, Adj. $R^2 = .293$, S.E. $= .76218$, Durbin-Watson $= 1.772$

The adjusted R^2 for model 2 was .293, indicating that the independent variables predicted approximately twenty-nine percent of the variation in penalty departures. The results from model 2 indicate that voluntary disclosure was not statistically significant, but that number of violations was significant ($p = .000$). In the first model, percent minority was not statistically significant. However, when percent African-American was included in the model and separated from percent Hispanic, the variable was statistically significant ($p = .006$).

Table 34 presents the results of the third regression model predicting the percent of departure in penalty amount. Model three included percent Hispanic within a three-mile radius and excluded percent African-America within a three-mile radius to determine if there are different ethnicity effects. The regression equation included the following independent variables: number of violations (vionum), voluntary disclosure (voldis), and percent Hispanic within a three mile radius of the violating facility (hisper3).

Table 34: Regression Model 3 Predicting Penalty Difference

Variable	B	SE b	Beta	VIF	t	Sig. T*
hisper3	2.835	.005	.067	1.021	.577	.566
vionum	.444	.109	.474	1.021	4.089	.000*
voldis	-1.257	.821	-.002	1.006	-.015	.988

$p<.05$, $R^2 = .238$, Adj. $R^2 = .198$, S.E. $= .81200$, Durbin-Watson $= 1.772$

The adjusted R^2 for model 3 was .198, indicating that the independent variables predicted approximately twenty percent of the variation in penalty departures. The results from model 3 indicate that again, voluntary disclosure was not statistically significant, while the number of violations was statistically significant ($p = .000$). Unlike percent African-American, percent Hispanic within a three-mile radius had no significant effect on the penalty difference ($p = .556$).

Table 35 presents the results of the fourth regression model predicting the percent of departure in penalty amount. The fourth regression model included percent below poverty in a three-mile radius to determine if there are socioeconomic effects. The regression equation included the following independent variables: number of violations (vionum), voluntary disclosure (voldis), percent African-American within a three mile radius of the violating facility (aaper3) and percent below poverty within a three-mile radius (bppov3).

Table 35: Regression Model 4 Predicting Penalty Difference

Variable	B	SE b	Beta	VIF	t	Sig. T*
aaper3	-1.304	.005	-.315	1.478	-2.385	.020*
bpper3	4.686	.015	.004	1.455	.032	.974
voldis	-.151	.780	-.021	1.014	-.194	.847
vionum	.406	.106	.433	1.075	3.847	.000*

$p<.05$, $R^2 = .328$, Adj. $R^2 = .281$,

S.E. = .76883, Durbin-Watson = 1.775

The R^2 for regression model 4 presented in Table 35 was .328. The adjusted R^2 was .281, indicating that the independent variables predicted approximately twenty-eight percent of the variation in penalty departures. The results from model 4 again show that voluntary disclosure is not statistically significant, number of violations is statistically significant ($p = .000$), percent African-American is statistically significant ($p = .020$), and that the percent below poverty in a three-mile radius had no significant effect on the penalty difference ($p = .974$).

Table 36 presents the results of the fifth regression model predicting the percent of departure in penalty amount. The fifth regression model included two additional variables: Supplemental Environmental Project amount and Clean Air Act (CAA) violator. The regression equation included the following independent variables: number of violations (vionum), voluntary disclosure (voldis), percent African-American within a three mile radius of the violating facility (aaper3), SEP amount (sepamt), and CAA violator (CAA1).

Table 36: Regression Model 5 Predicting Penalty Difference

Variable	B	SE b	Beta	VIF	t	Sig. T*
sepamt	1.901	.000	.274	1.094	2.529	.014*
voldis	6.915	.733	.010	1.029	.094	.925
vionum	.416	.123	.359	1.050	3.387	.001*
aaper3	-9.214	.005	-.227	1.147	-2.044	.046*
caa1	.262	.201	.148	1.195	1.307	.197

$p < .05$, $R^2 = .410$, Adj. $R^2 = .357$

S.E. = .71662, Durbin-Watson = 1.764

The R^2 for the fifth regression model presented in Table 36 was .410. The adjusted R^2 was .357, indicating that the independent variables predicted approximately thirty-six percent of the variation in penalty departures. The results from model 5 indicate that the best predictors of penalty difference are the number of violations committed ($p = .001$), the Supplemental Environmental Project amount ($p = .014$), and the percent African-American in a three-mile radius ($p = .046$).

Summary of Results for Research Question 4

The regression models indicated that percent African-American population was a significant predictor of penalty departure amounts controlling for the effects of the number of violations, supplemental project amount, and whether or not violations were voluntarily disclosed. The various regression models estimated above indicated that ethnicity population effects had a marginal to weak (statistically insignificant) effect, and that class-effects were not evident. Overall, however, the models predicted only a modest amount of variation in penalty difference amounts, and it is plausible that omitted factors may explain this variation.

Although the relationship between African-American population and penalty departure is significant, it is difficult to interpret the regression coefficients because penalty differences were logged. Roughly, a one-percent increase in the African-American population decreased penalty departure amounts by more than 9 percent. To clarify the relationship between race and penalty amounts further, the distribution of penalties for communities at opposite ends of the racial composition spectrum were also examined (see Table 37). For example, for the 8 communities that were 60 percent or more African American, 5 involved no supplemental amount. In addition, while 6 of these eight communities experienced a penalty departure, the departures were less than $255,000 in all cases. The average penalty sought by the EPA in high concentration African-American communities was $171,108, while the average penalty difference was $13,608.

Table 37: EPA Requested Penalty, Penalty Differences and Supplemental Penalty Amounts Across High Concentration African American, White, Hispanic, and Minority Communities.

	Requested Penalty	Penalty Difference	Supplemental Penalty	N
ALL	588,464	165,838	237,651	107
>60% AA	171,108	13,608	3,571	8
>60% WH	704,662	274,280	436,975	54
>60% HS	208,160	224,580	5,340	5
>70% MN	185,017	121,855	12,277	15

AA = African American; WH = White;
HS = Hispanic; MN = Minority (African American and Hispanic)

Comparing these results to those for communities that were primarily white (more than 60 percent white), fourteen of the 54 white communities (26%) received supplemental amount settlements. The mean supplemental amount was in excess of $436,000, or one-hundred and twenty-two times higher than the mean supplemental amount in high concentration African-American communities. In addition, 26 of the 54 white communities received penalty departures. In one case, the departure led to a lower fine than sought by the EPA ($ 23,850 less than the recommendation of $66,000). In four cases, the departures were in excess of $ 1,000,000. The mean departure for primarily white communities was nearly $275,000, or nearly twenty times higher than in primarily African-American communities.

These results support those from the regression models, though they do not take case seriousness into account. Both the supplemental and regression analyses results should be considered with caution given the small number of cases representing high African-American concentration communities.

Summary of Results

The results of the four research questions analyzed in the present study indicate that petroleum industry violations are widespread; media reporting of petroleum refining industry violations is limited; the seriousness of the offense based on penalty assessments is the greatest factor leading to coverage; and that penalty departures were lower in predominantly African-American communities. A discussion of these results is presented in Chapter Seven.

CHAPTER 7

Discussion

INTRODUCTION

The following chapter provides a discussion of the results reported in Chapter Six. Overall, the present study found that environmental crime in the form of federal petroleum refining violations is widespread; media coverage of petroleum refining violations is lacking; certain factors contribute to the likelihood of news coverage; and that penalty departures are affected by community racial characteristics.

Research Question #1

The purpose of Research Question #1 was to determine the nature and distribution of environmental crime as indicated by federal petroleum violations. Several variables were examined and the results for each variable were reported on in Chapter Six. The results for a number of the descriptive statistics deserves further discussion. The key findings were as follows:

- **Eighty-three percent** of the total number of petroleum refining companies in the United States were responsible for one or more EPA violations from 2001-2002.

- **Thirty percent** of our nation's petroleum refining companies were involved in two or more EPA cases from 2001-2002.

- **Thirty percent** of the cases involved more than one federal environmental statute violation.

- Petroleum refining violations occurred in **eighty-eight percent** of states hosting petroleum refining operations.

- **One out of every five EPA cases** involved a serious violation of environmental statutes.

- Only a small proportion of cases **(13%)** were the result of facility self-disclosures.

- From 2003-2004, the EPA **did not** inspect 54.3% of facilities for violations of the CAA; the EPA **did not** inspect 49.6% of facilities for violations of the CWA; and the EPA **did not** inspect 48.4% of facilities for violations of RCRA.

- From 2003-2004, states **did not** inspect 31.3% of facilities for violations of the CAA; states **did not** inspect 53.1% of facilities for violations of the CWA; and states **did not** inspect 49.2% of facilities for violations of RCRA.

- Approximately half (48.4%) of all facilities were in **non-compliance** of the CAA for 7 or 8 quarters (of 8) from 2003-2004.

Company Information

The Environmental Protection Agency filed and/or settled 162 cases against petroleum refining companies from 2001-2002. Seventy-eight separate companies were involved in the 162 cases. According to the Department of Energy (2003), ninety-four companies own and operate the one hundred and forty-nine petroleum refineries located in the United States. Based on the data, *eighty-three percent of the total number of petroleum refining companies in the United States were responsible for one or more EPA violations from 2001-2002.* Sixteen of these companies were involved in three or more EPA cases, which represents 17% of the total number of petroleum refining companies operating in the United States. Twelve of these companies were involved in two EPA cases, which represents 13% of the total number of petroleum refining companies operating in the United States. *Thirty percent of our nation's petroleum refining companies were involved in two or more EPA cases from 2001-2002.* Furthermore, *thirty percent of cases involved violations of more than one federal environmental statute.*

These numbers suggest that federal environmental violations committed by petroleum refining companies are widespread and frequent. For example, Koch Industries was involved in ten cases,

Chevron was involved in nine cases, and Shell Oil was involved in eight cases. Furthermore, the data analyzed in the present study is reflective of violations *known* to the EPA. If these 162 cases were reflective of all petroleum refining industry violations, it would be fair to say that the petroleum refining industry is responsible for a great deal of environmental crime. However, these cases more than likely only represent a small proportion of the total amount of environmental crimes committed by the petroleum refining industry. Criminologically, this rate of offending would qualify these companies as persistent criminal offenders, and perhaps as career criminals. Future research should compare the nature and distribution of environmental crime indicated by data analyzed in the present study with data for other years in order to provide a more accurate picture of the trends in petroleum refining industry violations.

<u>Refinery Location</u>

Violating refineries were located in thirty states across the country. According to the Department of Energy (2003), thirty-four of our nation's states are home to one or more petroleum refineries. Based on the data, *petroleum refining violations occurred in 88% of the states hosting petroleum refining operations.* Refineries located in Texas, Oklahoma, Pennsylvania, Delaware, and Illinois comprise 24.5% of the total number of refineries operating in the United States; yet refineries located in just these five states were responsible for over half (51.8%) of all federal petroleum refining industry violations.

- Texas refineries comprise 16.8% of the total number of refineries in the United States but were involved in 25.9% of federal petroleum refining cases.

- Oklahoma refineries comprise 2.5% of the total number of refineries in the United States but were involved in 8.6% of federal petroleum refining cases.

- Pennsylvania refineries comprise 2.5% of the total number of refineries in the United States but were involved in 6.8% of the federal petroleum refining cases.

- Delaware refineries comprise less than one percent of the total number of refineries in the United States but were involved in 4.9% of federal petroleum refining cases.

- Illinois refineries comprise 2.1% of refineries in the United States but were involved in 5.6% of federal petroleum refining cases.

While the present study did not focus on EPA enforcement decisions, it is worth noting that certain states appear to be responsible for more than their fair share of federal petroleum refining violations. It may be that refineries located in these states are in fact violating federal environmental statutes with more frequency than refineries located in other states. However, enforcement decisions may play a role in the unequal distribution of federal enforcement actions by state. Future research should examine the distribution of federal petroleum refining violations by EPA region, state, and related variables. In addition, an examination of state environmental enforcement initiatives and trends may shed some light on why particular states are disproportionately targeted by the EPA. If certain states consider environmental statute enforcement to be a high priority, the EPA may be less likely to get involved. Conversely, if state environmental enforcement actions are lacking, the EPA may pay more attention to refineries located in these states.

Case Type

According to the EPA, administrative action should be taken unless: 1) the total penalty amount is in excess of $200,000; or 2) the offense was committed more than a year prior to the case issuance or 3) the nature of the violation requires injunctive relief or involves evidence of a criminal violation. Criminal and civil judicial cases then involve more serious violations.

Most of the cases in this sample (78.4%) were resolved by the EPA through administrative actions (the sample does not include criminal violations, which fall under the authority of the Department of Justice). However, in 21.6% of the cases (N=35), civil judicial action was required. The data indicate that *one out of every five EPA cases involved a serious violation of environmental statutes.* In order to get a better understanding of how serious judicial cases are in comparison with administrative cases, future research should examine violation details outlined in the ECHO database. Furthermore, future research should also include attempts to examine criminal charges brought against oil refineries. Existing data sources, however, do not allow researchers to access corporate identities from criminal case data, and special access to EPA data would be required to undertake such an investigation.

Voluntary Disclosure

In December of 1995, the EPA introduced a program entitled "Incentives for Self-Policing: Discovery, Disclosure, Correction, and Prevention of Violations". The policy was designed to provide major incentives for companies that voluntarily discover, promptly disclose, and expeditiously correct noncompliance with environmental statutes. This program was designed to give companies the opportunity for a reduction in fines if they voluntarily disclosed existing violations. According to the EPA, violations can be considered voluntarily disclosed even if they are discovered during the course of an environmental audit. In the present study, *only a small proportion of cases (21 or 13%) were the result of facility self-disclosures.* The EPA insists that since the program was initiated "an increasing number of companies have voluntarily come forward to disclose environmental violations". However, the EPA does not provide the actual statistics for the impact of the policy on self-disclosures. Future research should examine the trends in self-disclosures and calculate the amount of penalties that would have been assessed had the self-disclosure incentives not existed.

The voluntary disclosure policy led to fine reductions in all cases where the violation was voluntarily disclosed. Consequently, the seriousness of the violations based strictly on penalty amounts may be affected by this policy provision.

EPA and State Inspections

From 2003-2004, the EPA did not inspect 54.3% of facilities for violations of the CAA; the EPA did not inspect 49.6% of facilities for violations of the CWA; and the EPA did not inspect 48.4% of facilities for violations of RCRA. From 2003-2004, states did not inspect 31.3% of facilities for violations of the CAA; states did not inspect 53.1% of facilities for violations of the CWA; and states did not inspect 49.2% of facilities for violations of RCRA. Approximately half of all facilities included in the present study were not inspected by the EPA in the most recent two-year time period. Approximately half of all facilities were not inspected by the state for violations of the CWA and RCRA and one-third of all facilities were not inspected for violations of the CAA. These findings suggest that violations might be considerably higher were the EPA and/or the state to inspect all facilities for petroleum refining violations. The reasons for lack of inspections are probably diverse and future research should seek to determine how the EPA and the states structures their inspection endeavors. If the EPA or the state has a systematic inspection format, it seems reasonable to suggest that facilities may aim for compliance when they are aware of an upcoming inspection. If, however, the EPA and the state ensures that inspections are random, the possibility of discovering violations will likely increase.

Non-Compliance of CAA, CWA, and RCRA

The EPA collects data on the number of quarters of non-compliance of the various environmental statutes. The data show that approximately half (48.4%) of all facilities were in non-compliance of the CAA for 7 or 8 quarters from 2003-2004. Fifteen facilities (11.7%) were in non-compliance of the CWA for 7 or 8 quarters from 2003-2004. Approximately one-third (27.3%) of facilities were in non-compliance of RCRA from 2003-2004. Overall, it appears that a significant proportion of facilities are often in non-compliance with federal environmental statutes, and are persistent, repeat offenders. The EPA does not undertake action against every facility that is frequently in non-compliance. If the EPA were to file cases against every facility for every violation, the EPA would need to increase its budget and enforcement staff tenfold.

Discussion Summary Research Question #1

Although the data show that environmental crimes committed by the petroleum refining industry are widespread and persistent, it appears that enforcement actions are not consistent with the amount of crime committed by the petroleum industry. More than likely the industry is aware that inspections are infrequent and that even in the case of consistent non-compliance, the EPA may not undertake enforcement actions. It appears that the petroleum refining industry would rather violate environmental statutes and risk being caught, rather than comply. If the risk of being caught is low and punishment is rare or involves minimal economic impact, then the financial incentives to violate the law are high. The results indicated that environmental crimes committed by the petroleum refining industry are widespread, even when one only takes into consideration *crimes known to the EPA.*

Research Question #2

The purpose of Research Question #2 was to determine the nature and distribution of mainstream news media reporting of federal petroleum industry violations. Seventy-four articles were collected which made reference to federal petroleum industry violations. However, of these seventy-four articles only seventeen articles (23%) corresponded directly with cases included in the EPA ECHO database. *The mainstream mass media does not focus much attention on federal petroleum refining violations.* The results from the present study add to the growing body of evidence which shows that media reporting of corporate crime and/or environmental crime is minimal (Maguire, 2002; Lynch, Stretesky, and Hammond; Lofquist, 1997; Wright, Cullen, and Blankenship, 1995; Lynch, Nalla, and Miller, 1989; Morash and Hale, 1987; Evans and Lundman, 1983; Swigert and Farrell, 1980). Future research that broadens the scope of the search parameters to include a wider range of years and cases may reveal a greater number of articles for analysis. Despite the low number of articles reporting on the petroleum refining industry, the data do have some noticeable characteristics which are discussed below.

News Source

Over 20.3% of the articles were reported in the *Houston Chronicle*, which is not surprising considering that Texas hosts more petroleum refineries (25) than any other state. The *Times Picayune* (New Orleans) was responsible for 16.2% of articles, which again, is not surprising as Louisiana hosts the third largest number (17) of the nation's petroleum refineries. What is most surprising is the number of articles produced by the *Star Tribune* in Minneapolis (7) was rather large, despite the fact that only two refineries are located in the entire state. *The New York* Times also contained seven articles pertaining to the petroleum refining industry despite the fact no petroleum refineries were operating in the state.

Overall, the data suggested that reporting decisions may be based on the importance of petroleum refining in the local community and on individual decision-making by reporters. For example, some reporters may feel that the petroleum refining industry is newsworthy while other reporters may disagree. Ultimately, editors must decide if the article is newsworthy. Future research should pay particular attention to news reporting decision-making processes in order to provide a better understanding of the process.

Article Location

News articles on the petroleum refining industry appeared in one of four locations: 1) News (Section A and National News), 2) Local News (Section B, Metro, or Suburban), 3) Business or Money, or 4) Other (including Science or Editorials). The location of the article is significant in that certain sections of newspapers are more widely read than other sections. According to a recent study by Mediamark Research Inc. (2004) on newspaper section readership, seventy percent of adults read the general news sections of the paper while only forty percent read the business or finance sections of the paper. Sixty-eight percent of petroleum refining industry articles appeared in the general or local news sections of the paper while twenty-eight percent of articles appeared in the business or money sections of the paper. Of the seventy-four articles, only eight articles appeared on the front page of the national news section. While sports and entertainment are given their own sections in the paper, the environment is reported on in various sub-sections of the paper.

Case Specific News Articles

Chapter Six reported on the content of newspaper articles directly related to cases included in the present study. The purpose of the following discussion is to present some of the themes apparent throughout the news articles that reported specifically on the cases included in the present study. Specific quotes taken from the various articles represent overall reporting trends observed in the seventeen case specific news articles. Discussions of the reporting trends are presented after each sample of article quotes.

Violations Committed

> "The sector (petroleum refining) is regarded as <u>one of the worst in terms of environmental compliance</u>" (*Times Picayune*, New Orleans, 7/26/00)

> "An Environmental Protection Agency investigation of U.S. refineries showed <u>widespread violations of the Clean Air Act,</u> with emissions problems from stacks, leaking valves, and other areas" (*Times Picayune*, New Orleans, 7/26/00)

> "The comprehensive agreement <u>frees the company from the threat of legal action by the government for past violation of clean air laws</u>" (*Times Picayune*, New Orleans, 7/26/00)

Only one article (*Times Picayune*, 2000) indicated that the petroleum refining industry is "one of the worst" violators of environmental statutes and that violations are "widespread", consequently the public is more than likely unaware of the extent to which the petroleum refining industry violates environmental laws. If more articles presented the petroleum refining industry as one of the worst violators in terms of environmental compliance and explained what that means in terms of environmental and human health consequences, perhaps the public would be more likely to support stricter environmental legislation and emissions standards. Furthermore, social activism may increase with the knowledge and understanding of the extent to which the refining industry violates environmental policies. The last quote states that the agreement in question will free the company "from the threat of legal action by the government for past violation of clean air laws". Similar statements appeared in other articles.

> "Fines cover pollution problems <u>involving air, water, and solid and hazardous waste</u> during the past 26 months" (*Star Tribune*, Minneapolis, 7/26/00)

> "The complaint against Clark lists violations of federal statutes, including the <u>Clean Air Act and the Clean Water Act</u>". (*St. Louis Dispatch*, 9/10/98)

"The violations have <u>included illegal dumping, illegal emissions, falsifying environmental reports, wetlands destruction, sewage overflow, chemical discharges, and oil spills</u>" (*Washington Post*, 9/10/98)

"The wide-ranging Norco probe centered on the company's (Motiva) <u>failure to monitor, check, and fix thousands of toxic leaks and on whether the company misrepresented its operations </u>to the agency (EPA)." (*Times Picayune*, New Orleans, 3/22/01)

Specific violations committed by the petroleum refining industry were, for the most part, reported in general terms. For example, *The Star Tribune* (2000) described the violations as "pollution problems involving air, water, and solid and hazardous waste". *The St. Louis Dispatch* (1998) (as well as other articles) listed the environmental acts that were violated. *The Washington Post* (1998) and the *Times Picayune* (2001) can be credited to some degree for providing a more detailed and specific list of the violations committed by the petroleum refining industry. Reporting violations in vague terms with sweeping generalizations does not present a clear picture of what is actually occurring. For example, stating that a company violated the Clean Air Act, without a precise description of how and what impact the violation has on the local community, has an effect on how the public perceives the violations. If reporters provided more detailed descriptions of the violations and environmental and human health consequences, then perhaps the public would be more willing to view environmental offenses as criminal offenses.

Environmental and Human Health Impacts

"Toxins include benzene, a <u>known carcinogen</u>, and <u>smog-causing compounds</u> such as nitrogen oxides, sulfur dioxides, and volatile organic compounds" (*Star Tribune*, Minneapolis, 7/26/00)

"The pollutants in question are <u>smog-forming</u> nitrogen oxide; chemical gases that form smog, including <u>cancer-causing</u> benzene; sulfur dioxide, and tiny soot particles." (*Houston Chronicle* 3/22/01)

"Air pollution triggers such illnesses as <u>childhood asthma and cancer</u>" (*Star Tribune*, Minneapolis, 7/26/00)

"The new equipment is intended to help ease <u>respiratory problems such as childhood asthma</u> by cutting pollutants such as nitrogen oxides, sulfur dioxide, particulate emissions, carbon monoxide, benzene, and volatile organic compounds" (*Houston Chronicle*, 5/12/01)

News articles in the *Houston Chronicle* (2001) and *Star Tribune* (2000) did a fairly good job of listing some of the major pollutants released into the air as a result of violations. However, although many people know that sulfur dioxide and carbon monoxide are harmful substances, the dangers associated with other substances are not as apparent. The articles state that these pollutants are associated with respiratory problems and are known cancer-causing agents. The environmental and human health consequences of exposure to there pollutants are much more serious and vast than indicated in these articles. For example, the *Star Tribune* (2000) states that "air pollution triggers such illnesses as childhood asthma and cancer". This statement underplays the seriousness of air pollution with the use of the word "triggers". Furthermore, cancer is a much more serious illness than childhood asthma. The spectrum of human health consequences are not explored in these articles which means that the public will remain ignorant to the problems caused by pollutants especially in light of the fact most articles did not mention environmental or human health problems at all.

"In a statement, <u>Norco refinery Manager Allen Kirkley</u> said he was confident the deal would help the environment and prove beneficial to the residents of the community of Norco." (*Times Picayune*, New Orleans, 3/22/01)

"The projects include $280,000 to finance a <u>cancer study of the effects of industrial chemical exposure</u> in the lower Mississippi River" (*Times Picayune*, New Orleans, 3/22/01)

The two quotes presented above, if taken at face value, may not be cause for concern. However, if one reads between the lines, several questions arise. In the first statement, a Norco refinery manager stresses that the enforcement "deal" will "help the environment and prove beneficial to residents of the community of Norco". If Norco was truly concerned about the environment and the local community, why did they violate environmental laws in the first place and only make a "deal" with enforcement officials when they were caught? Why does the article present a Norco representative commenting on how the company (the offender) is benefiting the environment and the community (the victims)? Where are the voices of the victims? The second quote refers to the funding of a cancer study to analyze the effects of industrial chemical exposure. Why do we wait for problems to arise before we study them? We have known for a long time that chemical exposure is linked with cancer and other illnesses. Studies of this type take years to complete. In the meantime, people will continue to get sick due to chemical exposure.

> "Residents of nearby Park Hill criticized the length of time the company (Conoco) was allowed to delay adding (air pollution) controls. 'I think we have to look at what these emissions will cost in terms of people's health', said Roz Wheeler-Bell, 50, chairwoman of Greater Park Hill Community, Inc. 'I don't think improvements six or seven years down the road is any great victory.' Wheeler-Bell said the neighborhood is often washed by a 'chemical, burny' odor that sets off her son's asthma and brings complaints of headaches from other residents. 'It just smells toxic', she said, 'And the consensus is it's gotten worse in the last few years'." (Denver Post, 12/21/01)

Only two articles of the seventeen case specific articles actually quoted a resident (victim) from the local community and in both articles (*Denver Post*, 2001; *Rocky Mountain News*, 2001), the same resident was quoted. While the first part of the articles contained quotes from EPA officials and the violating company, which included words and phrases such as "agreement", "settlement", and "victory for everyone involved", the comments made by resident, Roz Wheeler-Bell painted a different picture. She emphasized that the so-called "victory" was not a victory for local residents. While an enforcement decision did penalize Conoco, the company was given several years to fix the problem. And in the meantime, residents continue to suffer numerous health consequences of exposure to toxic air. No other case specific article addressed resident concerns. It would appear that the victims have been forgotten when newspapers report on petroleum refinery violations.

Actions Speak Louder Than Words

"Protecting our natural resources through strong enforcement of environmental law is a top priority for the Department of Justice" Attorney General John Ashcroft. (*Times Picayune,* New Orleans, 3/22/01)

"Edward L. Dowd Jr., the U.S. attorney for the Missouri District that includes St. Louis, is one of the nation's most aggressive prosecutors of waterway crimes, with 10 convictions in the last year. 'We're sending a message that you can't pay a fine and walk away,' Dowd said in an interview. 'This is serious stuff. We're going to make you repair the damage you've done. And if you did it on purpose, we're going to send you to jail." *Two sentences later.* . ."Environmental crimes are still a tiny faction of the (Justice) department's work, much less than 1 percent of its overall prosecutions" (*Washington Post*, 9/10/98)

"Officials who gathered at the river's banks yesterday said they were determined to protect the Mississippi from daily dumpings of raw sewage, cyanide, slaughterhouse waste, heavy metals, and other toxic junk" (*Washington Post*, 9/10/98)

"Attorney General Janet Reno (picture included with her quote) "vowed to hunt down all polluters 'in all corners of the watershed' and that they would not be let off with mere fines and apologies. However, Lois J. Schiffer, assistant attorney general for environment and natural resources, said the office has investigated criminal proceedings against Shell and decided that a 'strong civil settlement was our only logical course in this case'". (*St. Louis Dispatch*, 9/10/98)

Many of the case specific articles presented quotes from high-ranking state and federal officials. It comes as no surprise that these officials were adamant about going after violators. But actions speak louder than words. According to Attorney General John Ashcroft (Times Picayune, 2001), "strong enforcement of environmental law is a top priority of the Justice Department". Similarly, Edward L. Dowd Jr. stated in the Washington Post (1998), "This is serious stuff. We're going to make you repair the damage you've done. And if you did it on purpose, we're going to send you to jail." However, in the same article, just two sentences later, the following statement appears, "environmental crimes are still a tiny faction of the (Justice) department's work, much less than 1 percent of its overall prosecutions". Attorney General Janet Reno was quoted in the St. Louis Dispatch (1998), stressing that polluters "would not be let off with mere fines and apologies". But, just one sentence later, the truth is revealed, "Lois J. Schiffer, assistant attorney general for environment and natural resources, said the office has investigated criminal proceedings against Shell and decided that a 'strong civil settlement was our only logical course in this case'". It appears that officials are overzealous with their words. Environmental crime enforcement is NOT a major priority for the Justice Department. Quotes from officials make it sound like our federal and state agencies are deeply concerned about the problems caused by the petroleum refining industry. These quotes may even lead the public to believe state and federal agencies are committed to serious enforcement of our environmental laws. If so, the public is misguided. Enforcement of environmental laws is not given the serious attention indicated by our nation's leaders.

> "The Motiva-Norco case demonstrated the state's ineffectiveness in policing its major industrial sites. . .a DEQ inspector did not find any violations of the leak-detection program shortly before (a former employee) came forward. 'These violations would not have surfaced had it not been for someone stepping out and effectively dragging the agencies to the violations', Mark Davis, executive director of the Coalition to Restore Coastal Louisiana.(*Times Picayune*, New Orleans, 3/22/01)

Only one article reported on problems with environmental enforcement (Times Picayune, 2001) and even so, the emphasis was on problems with enforcement at the state level. While the present study did not focus directly on federal and state enforcement issues, it is worthy of discussion. The data indicate that violations are widespread, inspections are sporadic, and that enforcement is lacking. The quote above indicates that a state inspector found no evidence of violations. However, only a short time later, a former employee made the effort to inform agencies of existing violations. This article suggests that even when inspections are being conducted, violations are overlooked. Are state inspectors intentionally overlooking violations? Future research should examine state and federal enforcement efforts and processes. If state agencies and the EPA are ineffective enforcers of environmental legislation, we need to know about it so that changes can be made.

Company Portrayal

"Koch and BP Amoco came forward promptly to work on problems at their refineries" (EPA Administrator as quoted in the *Star Tribune*, Minneapolis, 7/26/00)

"After being alerted by government officials, the two companies (BP and Koch) initiated talks with the EPA in March to avoid a lawsuit." (*Times Picayune*, New Orleans, 7/26/00)

"This agreement represents a strong proactive environmental initiative by Koch, consistent with our proven commitment to environmental stewardship and other voluntary clear air initiatives" Jim Mahoney, a Koch Petroleum Group executive vice president. (*Times Picayune*, New Orleans, 7/26/00)

The quotes stated above highlight a theme prevalent throughout the case specific articles. The offenders are often praised for their quick response to EPA inquiries and articles portray the companies as willing to work on the problems they caused. The second quote stated above is rather noteworthy. It states that government officials alerted BP and Koch about violations at their facilities. Due to this knowledge, BP and Koch "initiated" talks with the EPA. It appears that companies are ready and willing to work out problems they caused, but if and only if, they are caught red-handed. Reporting presents the companies in favorable terms despite the seriousness of the violations committed by these companies. For example, Koch Industries was involved in 10 EPA cases from 2001-2002 (the most of any company) yet the article paints a different picture of Koch. By allowing the offender to comment (third quote above) and excluding victim accounts, the public is seriously misinformed as to the true nature of the violations and consequences. The offender says "I'm sorry (that I was caught)", the EPA says, "Okay, pay a fine and don't do it again", and the victims are ignored. The company then goes on to violate the law, again, and again, and again.

> "In a telephone press conference, Browner <u>praised BP and Koch for their cooperation</u> in bringing the cases to a speedy conclusion". *At the conclusion of the same article,* "In January, Koch agreed to pay $30 million in fines and spend $5 million on environmental projects for spilling an estimated 3 million gallons of oil from pipelines in Texas and five other states. Browner said at the time that <u>Koch negotiators had been stubbornly unwilling to accept responsibility</u> for the environmental damage caused by the spills". (*Houston Chronicle*, 7/26/00)

> "The companies <u>deny all the allegations</u>" *Three sentences later* "EPA Administrator Christie Whitman <u>praised the three companies</u> for 'taking the initiative to resolve their environmental problems cooperatively and quickly'". (*Houston Chronicle*, 3/22/01)

Similar to the quotes discussed in the previous paragraph, these two quotes from the Houston Chronicle (2000, 2001) report on the EPA's "praise" of BP and Koch's cooperation. However, in both articles, contradictory information is reported. In the first article from the Houston Chronicle, the EPA praises the companies for their cooperativeness but at the conclusion of the same article, it states that "Koch negotiators had been stubbornly unwilling to accept responsibility" for the damage they caused in the past. It appears that past behavior is unlikely to tarnish a company's reputation (despite the fact that this same company was involved in ten EPA cases in a two year time period). In the second article, it is reported that the companies "denied all the allegations", yet according to the EPA, the companies should be praised for "taking the initiative to resolve their environmental problems cooperatively and quickly". How are the companies able to resolve the problems cooperatively and quickly if the problems don't exist?

Evidence of Criminal or Negligent Activity

> "The company (Koch) agreed to pay a $6.9 million fine in 1998, primarily to the state of Minnesota, and $8 million last

fall to resolve a <u>federal criminal complaint</u>" (*Star Tribune*, Minneapolis, 7/26/00)

"A significant number of the issues surfaced when a <u>former Norco employee disclosed environmental problems to regulators.</u>" (*Times Picayune*, New Orleans, 3/22/01)

"<u>Criminal inspectors</u> have honed in on Motiva's record-keeping in Norco to determine if <u>the company falsified records.</u>" (*Times Picayune*, New Orleans, 3/22/01)

"The refinery was listed on the state Superfund site in 1987, and has been the subject of several previous legal settlements, including upgrades of its petroleum storage tanks after a major gasoline leak in 1994. The company also treated more than 330 million gallons of contaminated ground water and 14,000 cubic yards of contaminated soil during the mid-1990s, the result of a <u>half-century of petroleum spills and leaks from tanks and underground pipelines</u>" (*Star Tribune*, 5/12/01)

"The Shell case, for example, <u>did not lead to a criminal prosecution</u>, even though the company faced allegations of illegal levels of sulfur dioxide, hydrogen sulfide, and benzene emissions. (*Washington Post*, 9/10/98)

Evidence of criminal or negligent activity was minimally discussed in the case specific news articles. The quotes presented above are the only examples of criminal or negligent behavior on the part of the company. Most articles do not discuss the cause of the violations, or they create the false perception that the company was unaware that they were violating environmental statutes. "We had no idea", appears to be the company motto when it comes to environmental violations. The companies are not called "criminals", the environment and affected communities are not depicted as "victims". The violations and offenses are not labeled as "crimes".

Discussion Summary Research Question #2

In an ideal world, the news media would report on every violation committed by the petroleum refining industry or at the very least, on the most serious cases. Consequently, the public would be aware of the widespread nature and distribution of environmental crime. Increased knowledge could lead to increased efforts to do something about the problem. Instead, we live in a society in which news reporting of this type of crime appeared in just seventeen articles over a six year time period. It is no wonder the public perceives environmental crime as less serious as compared to street crime. On a daily basis, our national and local newspapers report on violent crimes from around the world. Months may go before we see a news article reporting on environmental crimes committed by the petroleum refining industry. And even then, what *is* reported may actually do more harm than good. Results support the findings from previous research which suggests the media construction of corporate and environmental crime fails to adequately represent the actual nature of such crime (Maguire, 2002; Lynch, Stretesky, and Hammond; Lofquist, 1997; Wright, Cullen, and Blankenship, 1995; Lynch, Nalla, and Miller, 1989; Morash and Hale, 1987; Evans and Lundman, 1983; Swigert and Farrell, 1980). In the present study, the news reporting was never critical of the industry or of the government's lack of enforcement efforts. The news reporting relied on EPA information and had no investigative component. The news reporting allowed the offenders to comment but ignored the victims. Overall, media reporting of federal petroleum refining industry violations is not only lacking; it is misleading and downplays the serious of environmental crime.

Key Findings from the Star-Telegram, 2004 (Fort Worth, Texas)

Although the present study did not analyze news article coverage beyond 2003, one recent article that appeared in the *Star-Telegram* deserves recognition. The information reported on in the article was based on a research project which analyzed EPA compliance and enforcement data on the petroleum refining industry; the same data analyzed in the present study. Investigative journalists and environmental reporters collected and analyzed the data. In addition, project members conducted numerous interviews and visited refineries in Texas, Louisiana, and Delaware. The presentation of such an article in the news media is especially poignant since articles addressing petroleum refining industry violations are virtually non-existent. Furthermore, the article appeared in the National news section and contained over 4,000 words. The findings from the *Star-Telegram* are similar to the findings reported in the present study but with the addition of more data, including qualitative data, several key statements are worthy of review.

• Comprehensive clean-air inspections, a crucial step in identifying violations, are down 52 percent for refineries since 2001, compared with 4 percent for all industries.

• Notices of violations have plummeted 68 percent for refineries, compared with a 24 percent drop for all industries. And formal enforcement actions are down 31 percent for refineries but less than 1 percent for all industries nationwide.

• Refineries' increased self-reporting of pollution data has in many cases replaced on-site inspections by government regulators, and the EPA does little or nothing to ensure that the companies' reports reflect reality.

• Texas and Louisiana are home to five of the nation's 10 worst offenders when it comes to toxic air pollutants from oil refineries. But the EPA regional office responsible for those states, along with Oklahoma, Arkansas and New Mexico, has no air inspectors dedicated to the region's 50 petroleum refineries. The office's 15 air inspectors are responsible for about 3,000 industrial facilities that the EPA has classified as "major" emission sources. Throughout Region 6, 60 full-scale air inspections were conducted at oil refineries in 2003 -- by far the fewest since at least 1984.

• BP's Texas City refinery, the nation's largest, emits more toxic pollution into the air than any other U.S. refinery. Yet it hasn't had a comprehensive air-quality inspection in nearly three years.

• In the Corpus Christi area, Valero's refinery hasn't had a full-scale inspection in three years and one month, the Koch Petroleum refinery in four years and two months.

• In other parts of the country, from Chevron's refinery in El Segundo, Calif., to Premcor Refining Group's plant in Delaware City, Del., refineries with long histories of violations have also gone years without full inspections.

• In Texas, which has more refineries than any other state, the Commission on Environmental Quality is responsible for inspections. But like the EPA, it has no air inspectors dedicated to refineries. Instead, the air inspectors, now down to 129 after steady declines since 1998, handle compliance for as many as 2,000 major industrial sources from Brownsville to Amarillo.

- Since late 2000, the EPA has signed consent decrees with 11 oil companies, covering 42 of the 145 operating U.S. refineries. The settlements set deadlines for the companies to pay nearly $40 million in penalties and install an estimated $1.9 billion in pollution controls, among other requirements. In return, companies are released from liability for past violations.

The Star-Telegram report is noteworthy on two major fronts; 1) the article is extremely critical of the petroleum refining industry and the EPA; and 2) it serves as an example of exceptional media reporting of environmental crime. The article discusses the impact of political and economic decision-making which priorities dollar amounts over human health concerns. Overall, the presence of such an article may either be an anomaly or perhaps a sign of good things (more environmental crime reporting) to come.

Research Question #3

The purpose of research question #3 was to determine which factors appeared to lead to greater news coverage of federal petroleum industry violations. As indicated in the results from Chapter Six and in the preceding discussion, very few EPA cases (17) receive any attention from the mainstream news media. Although newspaper coverage of federal petroleum refining violations is lacking, the news articles that were reported suggest that certain factors lead to the likelihood of greater news article coverage. These major factors include: 1) initial data EPA case was filed by the EPA; 2) location of the violating refinery; 3) large penalty, compliance, and SEP amounts; and 4) refining capacity. Results from a content analysis of these articles were presented in Chapter Six. The following sections provide a discussion of the reasons why these factors appear to lead to greater news coverage of petroleum refining industry violations.

Initial Date of Case Filed by the EPA

Twelve of the seventeen articles (70.6%) appeared in news sources on the same day the EPA case was filed or in the week immediately following the initial case filing. No articles appeared on the day the case was settled by the EPA or on days immediately following the settlement. The violation data used in the news articles is more than likely gathered by reporters from the EPA through the Compliance and Enforcement Newsroom press release web page. If the EPA only provides press releases on the date the case was initially filed, it seems obvious why articles are reported in conjunction with the case filing date rather than the case settlement date. However, it appears reporters do not attempt to report anything more than what the EPA provides in their press releases. Perhaps if the EPA were willing to provide more detailed information to the press regarding the violations, reporters would be more apt to provide more details in their articles. On the other hand, if reporters were interested in delving deeper into the violation cases, they would discover a great deal of information in available on the EPA web site. It appears likely that reporting is linked to press releases provided by the EPA. It is not possible to link news articles with past press releases because the EPA does not keep archives of past press releases prior to 2003. Future research should compare article information with information presented in EPA press releases.

Location of the Violating Refinery

News articles addressing federal petroleum industry violations were more likely to appear if a violating refinery was located in the community in which the news source was primarily distributed. Thirteen of the seventeen articles (76.5%) addressed violations committed by local refineries. While this factor may not be surprising, it suggests that some reporters/editors consider environmental violations more newsworthy than other reporters/editors. Even though most of these articles made connections with local refineries, only two articles addressed local concerns and interviewed residents. More importantly, it appears that environmental crime in the form of petroleum refining industry violations is considered a *local* concern, not a national concern. Unlike homicides and other violent crimes (which make national headlines), environmental crime is only viewed as a local problem and therefore, not worthy of national attention, despite the fact that there are more environmental offenses committed each year than homicides.

Penalty Amounts

In sixteen of the seventeen articles (94%), EPA penalty sought, compliance amount, and SEP amount were reported. The cases reported on in news articles were all ranked as the top cases in terms of the amount of the penalty sought, the compliance costs, and SEP costs. News articles only reported on cases in which the total EPA costs exceeded $20,000,000. Consequently, cases involving lower penalties, compliance costs, and SEP amounts did not receive any coverage. It appears that the summation of the penalty amount sought, compliance amount, and SEP amount is the greatest factor leading to news article coverage of petroleum refinery industry violations. The problem with reporting only the most costly cases means that the seriousness of the case is determined by the penalty amounts and not by other more important factors. For example, harm to the community is not reported by the press nor measured by the EPA. High compliance costs are often associated with court costs and may not be directly linked to the seriousness of the case. In other cases, self-disclosure of violations led to the assessment of lower penalties by the EPA. Only the seven most costly cases received news coverage. Serious cases that did not receive penalties in excess of $20,000,000 in fines did not receive coverage despite the fact that many of these violations are responsible for a wide range of environmental and human health problems. Reporting of federal petroleum refining violations is directly related to economic issues and not human health issues.

Refining Capacity

News coverage of petroleum industry violations is more likely to occur in the violating company is responsible for a high percentage of the total United States refining output. Bigger companies receive more media attention. While not surprising, these findings suggest that smaller companies in violation of environmental statutes are less likely to receive news coverage. Again, the seriousness of the offense is overlooked when reporting decisions are made.

Discussion Summary Research Question #3

The purpose of news articles is to provide the public with newsworthy information. With the respect to the petroleum refining industry, it appears that information regarding violations is newsworthy if the EPA provides a press release and reporters do not need to investigate further. In addition, if a local refinery is in violation of environmental statutes, newsworthiness increases at the local level. The greatest factor related to reporting violations has to do with the amount of fines imposed by the EPA, rather than with the seriousness of the offense, which supports the findings presented by Randall and Defillippi (1987). If the offense did not result in more than $20,000,000 in fines, it was not newsworthy. Lastly, if the refining capacity of the company was relatively insignificant, the violations were not newsworthy.

Research Question #4

The purpose of research question 4 was to examine whether or not petroleum refining industry penalty assessment decisions were affected by the racial and socioeconomic composition of the communities surrounding the violating facility and to ascertain whether other factors influenced EPA penalty assessment decisions. The results indicated that community race characteristics were not significant predictors of the preliminary or assessed penalty. Consequently, the present study examined whether or not community race, ethnic, and class characteristics impacted the departure from the EPA penalty recommendation. For example, if the penalty initially sought by the EPA was $100,000, would the final penalty assessment be lower (or closer to the original amount sought) in minority and low-income communities and higher (greater than the original amount) in predominantly white and higher income communities. Several regression models were estimated in order to answer the revised question.

Voluntary Disclosure

Whether or not a company voluntarily disclosed a violation had no significant impact on the penalty difference. Although the EPA indicates that they are more lenient with facilities that self-disclose violations, the penalty differences in the present study did not lend support for leniency. This result however must be interpreted with caution due to the low number of companies that self-disclosed violations (21 cases or 13%). Other factors may have influenced EPA decisions despite the self-disclosure of the violations. The impact of voluntary disclosures on penalty assessments needs to be further examined, with the inclusion of a wider range of cases.

Number of Violations

Not surprisingly, companies with more violations received higher departures in the penalty amounts. In other words, the final penalty assessment was greater in cases involving multiple law violations. In cases involving just one violation, the final penalty assessment more closely matched the amount initially sought by the EPA. These results suggest that the EPA considers the number of law violations when making penalty assessment decisions.

Percent Minority, African-American, Hispanic, and Below Poverty

The percent minority (African-American and Hispanic), the percent Hispanic, and the percent below poverty within a three-mile radius of the violating facility had no significant impact on the penalty difference. However, percent African-American within a three-mile radius did have a significant effect on the penalty difference. As indicated in Chapter Six, a one-percent increase in the African-American population decreased penalty departure amounts by more than 9 percent. Results from the present study indicate that the percentage of African-American residents in a community surrounding a violating facility appears to have an effect on penalty assessment decisions. Primarily African-American communities are again, negatively and disproportionately impacted by environmental decisions which further supports the growing literature on environmental justice.

Supplemental Environmental Project Amount

Of the 162 cases included in the present study, eight cases involved violating facilities located in communities with a greater than 60 percent African-American composition while fourteen cases involved violating facilities located in communities with a greater than 60 percent White composition. The average supplemental environmental project amount in primarily African-American communities was $3,571. The average supplemental environmental project amount in primarily white communities was $436,000 or one-hundred and twenty-two times higher than the mean supplemental amount in primarily African-American communities. Furthermore, the mean departure for primarily White communities was nearly $275,000, nearly twenty times higher than in primarily African-American communities ($13,608). These results indicate that the racial composition of the community surrounding a violating facility have an impact of the penalty assessment decision-making process. However, due to the small number of cases involving high concentrations of African-American communities, these results must be interpreted with caution. These results however, do lend support for the underlying basis of the environmental justice movement which states that environmental hazards are disproportionately impacting African-American communities (Mohai and Bryant, 1992; UCC, 1987; Bullard, 1983) and that African-American communities receive less protection than predominantly white communities (Lynch, Stretesky, and Burns, 2004a, 2004b; Lavelle and Coyle, 1992).

Discussion Summary

The results from the present study contribute to the growing body of literature on environmental crime and justice and in particular, media coverage of environmental crime. To a large extent, the results presented in the preceding chapter are not surprising. Environmental crime in the form of petroleum refining industry violations is rampant. Almost every petroleum refining facility is in violation of one or more environmental statutes. Almost half of all facilities were in non-compliance of the Clean Air Act for 7-8 quarters of 8. As indicated by the infrequency of EPA and state inspections, the EPA and states do not appear to have the financial or enforcement personnel to effectively inspect every facility in a two year time period. Most companies do not self-disclose violations even though the EPA has specifically stated it will be more lenient with companies that self-report violations. When compared to primarily White communities, African-American communities are negatively and disproportionately impacted by penalty assessment decisions. Again, results indicate that environmental crimes committed by the petroleum refining industry are widespread, just taking into consideration *crimes known to the EPA*.

Although environmental crimes committed by the petroleum refining industry are widespread, the mainstream mass media does not focus much attention on this type of crime. Over a six year time period, only seventeen articles from the nation's leading newspapers reported on federal violations committed by the petroleum refining industry. Violations were rarely newsworthy unless they affected a local community and received fines in excess of $20,000,000. A content analysis of the seventeen articles reveals that what *is* reported may actually do more harm than good. There is no critical reporting on the industry as a whole. Environmental violations are reported in general terms. Human health consequences associated with the violations were either ignored or glossed over. Enforcement action was given a lot of hype in terms of quotes from EPA and government officials but no follow-up articles actually reported on enforcement results. The offenders are given more than enough press coverage and the ability to defend their actions, creating the image of a remorseful company that didn't realize they were in violation of an environmental statute rather than creating the image of a knowingly negligent offender committing an environmental crime. Only two of the seventeen articles included quotes from the actual victims of the environmental violations. If the public is only exposed to environmental crimes committed by the petroleum refining industry via newspaper reports, it should come as no surprise that these violations are not regarded as serious crimes, if crimes at all.

CHAPTER 8

Conclusions

INTRODUCTION

"When it comes to hazards in the workplace and the environment, the safe response, which has come to be accepted as scientifically responsible, is to say nothing and do nothing until we have clear proof that the hazard has actually made people sick" (Davis, 2002: xvii).

"We wasted fifty years debating the role of cigarettes in causing cancer, and we cannot afford to waste another fifty years before we develop strategies to prevent environmental cancer and other avoidable diseases" (Gaynor, 2002: x).

The purpose of this research was to highlight the importance of studying environmental crime and to examine media coverage of environmental crime. The purpose of this final chapter is to; present the various challenges facing environmental researchers and activists and to offer solutions; offer suggestions for environmental legislation and enforcement initiatives; provide an overview of corporate and environmental crime research; introduce sources of environmental crime data; emphasize areas for future inquiry and analyses; suggest ways in which environmental crime can be presented to the public through mainstream media outlets as well as educational endeavors; and suggest a social justice approach to dealing with environmental crime.

Challenges to Environmental Crime Research and Social Activism

Researchers engaging in environmental crime research and activists fighting for the environment and public health must take into consideration several challenges inherent in these pursuits. For one, the immediate consequences of an environmental offense may not appear obvious or severe. Consequently, environmental crime does not fit most people's perceptions of crime. Environmental crime researchers must, in addition to investigating and analyzing environmental crime data, provide rationalizations for their efforts. Furthermore, environmental crime is a complex issue that involves a wide range of problems, analyses, and so on.

> "Environmental criminology deals with concerns across a wide range of environments (e.g., land, air, water) and issues (e.g., fishing, pollution, toxic waste). It involves conceptual analysis as well as practical intervention on many fronts, and includes multi-disciplinary strategic assessment (e.g., economic, legal, social, and ecological evaluations). It involves the undertaking of organizational analysis, as well as investigation of monitoring, assessment, enforcement, and education regarding environmental protection and regulation. Analysis needs to be conscious of local, regional, national, and global domains and how activities in each of these overlap. It likewise requires cognizance of the direct and indirect, and immediate and long-term consequences of environmentally insensitive social practices" (White, 2003: 484).

Researchers must be prepared for the complexities associated with environmental crime research. Even if environmental crime researchers are successful in their empirical efforts, publishing may not be an easy task. Editors are often reluctant to publish studies which do not contribute to mainstream ideas of what constitutes crime. Activists must take into consideration the impact of corporate public relations campaigns which can influence the public's perceptions of corporate wrongdoing. Grassroots activists rarely have the financial resources with which to battle multi-million dollar "greenwash" campaigns, utilized by corporations to present an environmentally friendly image. The following section provides a more extensive discussion of the challenges facing researchers and activists and suggests ways in which these challenges can be overcome.

<u>Conceptualization of Crime</u>

Researchers are faced with a difficult uphill battle when they decide to research environmental crime. If, on the one hand, a researcher wanted to investigate homicide, s/he would not have to spend any time defending the selection of homicide as a crime, a serious infraction, or on definitions of what constitutes a homicide. On the other hand, environmental crime researchers not only have to define environmental crime, they must be prepared to describe in elaborate detail why environmental crime should be considered crime. The result is a theoretical and philosophical debate in the academic literature about the categorization and meaning of environmental crime. While environmental crime researchers are busy trying to "prove" environmental crime is indeed crime, corporations are continuing to illegally (and legally) release pollutants into the environment, and continuing to harm environmental and public health.

It is possible to change public conceptions about certain issues but not without extensive time and effort. For example, smoking cigarettes used to be considered socially acceptable. Political and public attitudes toward smoking have changed significantly since the 1960s, when the Surgeon General reported on the health hazards associated with smoking cigarettes. Numerous studies have shown that cigarette smoking causes various forms of cancer. Laws have been passed all across the nation that ban cigarette smoking in public venues. Anti-smoking campaigns have appeared in the mainstream mass media. But public attitudes toward smoking did not change over night. Even when it was extremely apparent that cigarette smoking caused numerous health problems, attitudes and behaviors toward smoking did not change immediately. We needed more proof and more studies to confirm that cigarette smoking was causing illness and disease.

Illustrating the laborious effort needed to prove cause and effect, Devra Davis, author of *When Smoke Ran Like Water: Tales of Environmental Deception and the Battle Against Pollution* (2002: xi-xii) tells a story about a study involving the air inside airplanes:

> "In the early 1980s, I reached a disturbing conclusion. I was working at the National Academy of Sciences on what turned

out to be a four-year-long study of air inside airplanes. The investigation didn't need to be four years, or even one. But Senator Daniel K. Inoyune had given the Federal Aviation Administration half a million dollars to fund a committee at the academy to find out why he kept getting sick after his regular eight-hour trips from Honolulu to Washington. . .I found out an easy way to answer the senator's question. From a friend at the Environmental Protection Agency, I borrowed a clunky piece of equipment called a piezobalance, which could measure the weight of airborne particles smaller than a human hair, such as those produced by cigarette smoke. I set off on a flight to Paris. . .By the end of the flight, I had the answer. The levels of particles in the smoking and nonsmoking sections were identical. The senator kept getting sick because for all his lungs cared, he might as well have been sitting with heavy smokers. When I got back to Washington I eagerly told my boss at the academy the good news. 'We don't need to do a study for the senator!' He looked at me nervously and asked, 'What are you talking about?' I suggested we could save time and money if we went out and studied a couple more planes and prepared a short report. After I explained what I had done, he sighed and shook his head. 'You can't do anything with those numbers. No committee reviewed what you were going to do. Nobody approved this project.' Half a million dollars and four years later, the official academy study confirmed what I had found in a single flight."

The purpose of Davis's story is to illustrate the importance of extensive research in our society. It is not enough to have a few studies linking environmental contaminants to human health problems. We need hundreds of valid and reliable studies to show that environmental pollution and toxins *cause* illness and disease. And even then, studies aren't enough. The information must be made available to the public in an understandable format. Most people do not read academic journals. Consequently, in order to change public perceptions as to what constitutes crime, information must be presented to the public in a "friendlier" format. For example, the movie Erin Brockovich had more of an impact on the public's perception of environmental crime than any academic study because it reached millions of people and also entertained as it informed.

Corporate and Political Backlash

Researchers and activists must take into account the vast amount of financial and political resources that corporations have to resist change. Corporations appear more willing to spend billions on touting an environmentally conscious image than to spend that money on actually changing their practices. "Greenwash" refers to public relations efforts to pose as friends of the environment through elaborate public campaigns. Corporations developed "greenwash" as a strategy for dealing with the community-based environmental movement. Corporate economists determined that it would be more cost effective to change the corporation's image rather than their practices. In the past few decades, corporations have *appeared* to become more concerned with the environment. They have established environmental departments with environmental personnel; created environmental programs; presented environmentally-themed media campaigns; and instituted voluntary policies and principles. On the surface, their efforts seem conscientious. However, a more careful examination reveals that safer alternatives are ignored; the worst polluters are the biggest supporters of Earth Day; and researchers and activists are treading in shark-infested waters:

> "In 2001, Elihu Richter and colleagues compiled a list of fourteen instances in the United States and other countries in which public health professionals were prevented from amassing data, where data were distorted by public relations concerns, where researchers were attacked or removed from their positions after warning of hazards, or where they were blackballed or gray-listed from participating in research on the environment" (Davis, 2002: 276).

What can researchers do? In order to counter corporate "greenwash", researchers and activists must first educate themselves on the tactics utilized by corporations. There are a number of excellent resources available that describe the PR strategies employed by corporations, and that detail ways of dealing with such campaigns (e.g., Greer, J. and Bruno, K. 1996. *Greenwash: the reality behind corporate environmentalism*: Apex Press and Beder, S. 2002. *Global Spin: the corporate assault on environmentalism*: Chelsea Green Publishing Company). Changing perceptions of what constitutes crime, simplifying environmental crime research for public consumption, and challenging corporate "greenwash" are not easy tasks. But they are necessary for the sake of our environment and health.

Suggestions for Environmental Legislation and Enforcement

One of the major criticisms of environmental legislation is its complex nature. One of the major criticisms of environmental enforcement is the lackadaisical effort given to criminal prosecution. While environmental crime prosecution has increased in recent years, there has been little progress in punishing or deterring corporate environmental violence. Regulatory agencies and the federal judiciary have been slow, cautious, or reluctant (if at all) to bring criminal charges against corporations for environmental crimes. In most cases of environmental crime, the probability of being caught is extremely low. While the present study cannot unravel all of the problems associated with environmental legislation and enforcement, a number of suggestions regarding legislation and enforcement can be suggested. Across our nation, at the federal, state, and local level, law enforcement officers, prosecutors, and judges are faced with an overwhelming task: to investigate, apprehend, and prosecute criminal offenders, in addition to a plethora of other responsibilities. Crimes such as homicide, rape, robbery, and assault are readily defined and victims/offenders more easily determined as compared to defining environmental crime and ascertaining victims/offenders. Resources and personnel are often scarce. Consequently, the majority of criminal enforcement and prosecution efforts center on traditional forms of crime. In order to effectively reduce environmental crime, we need to not only raise the punishment but also increase the probability of catching offenders. Environmental crime investigation and enforcement may be given more attention if:

- Law enforcement, prosecutors, and judges were made aware of the grave dangers associated with environmental crime.

- Law enforcement, prosecutors, and judges were given special training in understanding environmental legislation, investigation, and enforcement.

- Policies were instituted that required the pursuit of criminal penalties for environmental crimes rather than administrative or civil penalties. Administrative and civil remedies would only be used if there were no criminal remedies available.

- Individuals from multiple agencies gave their commitment to environmental crime enforcement and prosecution.

- Punishments for environmental crime included not only substantial fines and the threat of incarceration but also required mandatory clean-ups and publication of judgments in the mass media.

The threat of serious punishment, including incarceration, is a more effective deterrent than the threat of a fine. But "deterrence only works if the sanction to which the potential polluter is exposed is much higher than the amount of damage he might he causing" (Sjogren and Skough, 2004: 59).

Corporate and Environmental Crime Research

Recently, Lynch, McGurrin, and Fenwick (2004) examined the representation of white-collar crime and corporate crime research (1993-1997) in leading criminology journals, introductory criminal justice textbooks, and criminology Ph.D. programs. Lynch et al (2004) found that only 40 articles (3.6%) of 1,118 journal articles focused on white-collar or corporate crime and the majority (30 or 75%) appeared in two of the disciplines most critical or liberal journals. In each of the remaining mainstream journals, less than 2.3 percent of articles pertained to white-collar or corporate crime. Lynch et al (2004) examined 16 textbooks and found that of the 9,410 total pages of text, only 425 pages (4.5%) were devoted to white-collar or corporate crime. Only 9 of 21 Ph.D. programs offered a class in white-collar crime; however, most only offered the program once every two years and none of these programs required the course for completion of the doctorate. Lynch et al (2004) clearly show that white-collar and corporate crimes are seriously neglected in the criminological literature, in textbooks, and in course offerings. As a subset of these crimes, environmental crimes received very little attention.

Environmental Crime Data

It is obvious that our discipline needs an environmental awakening. Researchers complain that there is a lack of data available to study environmental crime. However, the complaints stem from the fact that there is not a centralized system of environmental crime data (Burns and Lynch, 2004). There is, however, a wealth of environmental crime data. Recently, Burns and Lynch (2004) published *Environmental Crime: A Sourcebook* which in addition to presenting an overview of environmental laws, the EPA, and enforcement practices, discusses in detail the various sources of environmental crime data which include data from EPA databases and non-EPA databases. The data *are* now readily available on the internet. In addition to the data described in Burns and Lynch (2004), there is a great deal of medical and epidemiological data that can be utilized to study human health and environmental crime. The following books also address ways in which environmental crime can be studied and analyzed:

> Murphy, B.L. and Morrison, R.D., eds. 2001. *Introduction to Environmental Forensics*. Academic Press.

> Gilbert, R.O. 1987. *Statistical Methods for Environmental Pollution Monitoring*. New York: John Wiley and Sons.

> Drielak, S.C. 1998. *Environmental Crime: Evidence gathering and investigative techniques*. CC Thomas.

Current Environmental Crime Research

Environmental crime research is a relatively new area. Within criminology, much of this research has been conducted within the past decade. Researchers have spent a great deal of time defining environmental crime and justifying its existence as a form of corporate violent crime. In addition, much of the environmental crime research has focused on issues related to environmental justice and the disproportionate impact of environmental hazards on minorities and the poor. The existing body of environmental crime and justice research has created a solid foundation for future research. Applied research is necessary in order to utilize the data to create meaningful change.

Future Environmental Crime Research

There is a wide range of future research endeavors relevant to environmental crime. According to Burns and Lynch (2004), the discipline of criminology needs to redirect the study of crime from street crimes to environmental and corporate crime.

Research should examine federal and state enforcement trends of environmental laws and compare federal/state enforcement of environmental laws with federal/state enforcement of other laws. Cross-cultural comparisons of environmental law, crime, justice, and enforcement actions as well as an examination of public perceptions and media coverage of environmental crime in other nations would greatly enhance the environmental crime literature. Furthermore, research should examine the role of the government as a major source of environmental pollution and examine the role of organized crime syndicates as sources of anti-environmental activities including the illegal disposal of toxic waste.

The research on media coverage of corporate and environmental crime remains largely limited. The research that has been conducted clearly shows that the media under-report corporate and environmental crime and do not treat corporate and environmental offenses as crime. Future research into the relationship between crime presentation and the media should take into consideration news media selection and production processes. In addition, research should examine the extent to which the news and entertainment media affect crime and justice attitudes, beliefs, and policies (Surette, 1992). The news media emphasizes extreme and dramatic cases. Environmental crime cases *are* extreme and dramatic stories. "Criminologists may well serve the commonwealth when they unmask the implicit biases of reporters and challenge the media to join the public discourse concerning the seriousness and potential criminality of corporate violence" (Wright et al., 1995: 35).

Although previous research as well as the present study contributes to the growing body of corporate and environmental literature, specific questions still need to be addressed and/or further examined. In January of 2000, David R. Simon presented a specific agenda for corporate and environmental crime research in the *American Behavioral Scientist* (Simon, 2000). Specifically, Simon suggested that the following questions be examined in future research endeavors (2000: 10-11):

1.   What additional violations of corporate crime laws are exhibited by the various chemical and other firms that have been convicted of multiple violations of hazardous waste and other environmental laws?

2.   What are the specific relationships between the firms convicted of numerous violations of various environmental laws and the EPA?

3.   What influence do powerful petrochemical and other firms frequently convicted of environmental criminal violations have on Congress and on the executive branch of the federal government?

4.   What patterns of criminality exist in which government agencies and corporations violate environmental laws in a co-conspiratorial fashion?

5.   How are victims of environmental crimes presented in the media? In addition, at what point does the mainstream mass media become concerned enough about environmental crimes to give them major and/or sustained attention?

6.   What corporate interlocks exist between firms in environmentally related fields and other sectors of American capitalism?

In addition to investigating these questions in future research, Simon (2000) suggests that it is necessary for criminologists to examine the relationship between environmental crime and major criminological theories. Future research should also continue to include the petroleum industry as a focus of inquiry. The petrochemical industry has a long history of criminal activity (Simon, 2000) and the EPA recognizes that the petroleum refining industry is one of the most polluting industries in the United States. Future analyses which include a larger sample of ECHO cases would more than likely lend greater support for the results reported in the present study.

Environmental Crime Activism

Future environmental crime research is a good step in the right direction. However, research isn't enough. Academics must be willing to become active participants in the fight to eliminate and reduce corporate and environmental violence. We must not only research, we must *act*. Beyond contributing to the growing body of environmental crime literature, criminologists can use their knowledge and expertise to educate the public and their students as well as to assist in grassroots activist efforts.

Newsmaking Criminology

> "Newsmaking criminology refers to criminologists' conscious efforts and activities in interpreting, influencing, or shaping the presentation of 'newsworthy' items about crime and justice. More specifically, a newsmaking criminology attempts to demystify images of crime and punishment by locating the mass media portrayals of incidences of 'serious' crimes in the context of all illegal and harmful activities; strives to affect public attitudes, thoughts, and discourses about crime and justice so as to facilitate a public policy of 'crime control' based on structural and historical analyses of institutional development; allows criminologists to come forth with their knowledge and to establish themselves as credible voices in the mass-mediated arena of policy formation; and asks of criminologists that they develop popularly based languages and technically based skills of communication for the purpose of participating in the mass-consumed ideology of crime and justice. A newsmaking criminology invites criminologists and others to become part of the mass-mediated production and consumption of 'serious' crime and crime control. It requires that they share their knowledge with the general public" (Barak, 1988: 566).

Academic involvement in the media process will not be an easy task. Chapter Two discussed the ownership of the majority of mass media outlets by just six conglomerates. Furthermore, many of the directors of these top media corporations sit on the boards of directors of some of the largest Fortune 500 companies and "interlock with each other through shared directorships in other firms" (Ruggiero and Sahulka, 1999). For example, NBC, Fox News, and Time Warner each have a board member who sits as a director on tobacco producer Philip Morris's board. According to Parenti (1997), "the Boards of Directors of print and broadcast news organizations are populated by representatives of Ford, G.E., G.M., General Dynamic, Coke, ITT,

IBM, Dow-Corning, Philip Morris, AT&T, and others. Given that distribution of ownership, it's not surprising that the concerns of labor are downplayed in the media". Criminologists can and should challenge the media elite and get involved in the media discourse on crime and justice.

To encourage the media to report on environmental crime, researchers with important information to present to the public need to establish ties with reporters and members of the press. MediaResource (mediaresource.org) is a non-profit organization which serves as a bridge between science and the media. According to MediaResource, journalists who contact the organization can get help at no charge in locating expert sources of information on science and technology to interview for their news and feature stories. MediaResource maintains a database of 30,000 scientists, engineers, physicians and policy-makers who have agreed to provide information on short notice to print and broadcast journalists. The Society of Environmental Journalists (www.sej.org) is also a good source of contact for criminologists who want to get involved in the media process. SEJ's primary goal is to advance public understanding of critically important environmental issues through more and better environmental journalism.

A survey of print and broadcast media journalists found that more than half of the journalists (52%) admitted they avoided stories that were too complex (Pew Research Center, 2000). Researchers must be able to explain their findings to reporters in clear and concise terms. In order to change public perceptions of crime, the mass media must not only report on environmental crime but also call it "crime". Researchers can assist with the reporting if they are willing to do the work. Tenure is based on peer-reviewed publications. More than likely, trying to get a message out to the public via the mass media will do nothing to further one's academic career. However, it will have a greater impact on our environment and health.

We know that the mass media play a critical role in the shaping of public perceptions of crime and justice. As criminologists, we need to become part of the social construction of public opinion of crime and justice. We cannot continue to leave the media construction of crime and justice solely to journalists and the media elite. In addition to establishing relationships with media personnel and providing different perspectives for crime news, criminologists can be more than just information sources. We can also *produce* crime information. Whether or not one agrees with the information presented by Michael Moore in his array of documentaries, it is apparent his efforts have attracted a great deal of public attention. Criminologists can work with media personnel to not only supply information but also to produce our own media displays of information in the form of documentaries, news briefs, and the like. Furthermore, criminologists can participate in community-based events and projects which unite local concerns, research agendas, and media attention.

Environmental Education

In order for environmental crime to be better understood and recognized as crime, it needs to be a topic that is taught to individuals of all ages. It wasn't until well into my criminology graduate program that I was even exposed to white-collar crime or corporate crime. Most colleges do not offer corporate or environmental crime courses. Most high schools do not even offer introductory criminology or criminal justice courses. Discussions of crime are generally included in social studies classes and are non-critical in nature. Children are told to watch out for strangers and to "Just Say No!"" I found just one book related to environmental crime aimed at middle and high school aged children (Arneson, D.J. 1991. *Toxic Cops*. Franklin Watts). There is very little research that examines criminal justice education at the elementary, middle, and high school level. This type of research is important. What are we teaching kids about crime? What we learn from the age of 5-18 has a major impact on what we believe as an adult. For example, I am constantly challenged by college students who have a difficult time accepting information I present even though it is based on academic research. Their opinions on certain topics were formed at a young age, by their parents, and were reinforced over fourteen years of school. When these same kids start college, they are dealing with huge life changes such as moving away from home for the first time, exposure to more choices, etc. College isn't just about learning, it is about adjustment. It is difficult for college professors to have an impact on student perceptions when we are competing against 1) parental opinions; 2) fourteen years of education; 3) and college adjustment issues. Consequently, criminal justice education should begin at a younger age.

According to Cheurprakobkit and Bartsch (2000), most public high schools do not offer criminal justice or criminology courses because they have difficulty finding qualified teachers and textbooks. Most instructors are law enforcement officers and the vast majority of textbooks are written for college students. General criminal justice education at the high school level is lacking; consequently exposure to environmental crime is not even on the radar.

Children may not need direct exposure to criminology and criminal justice education; however, they do need exposure to theories and perspectives that allow them to explore the relationship between humans and nature and our place in the natural world. Adults can benefit from environmental education as well. In general, science literacy in the United States is fairly low. According to Ross (1999), the public is overfed on information but starved for understanding. Knowledge of basic scientific concepts, facts, and vocabulary can make it easier for the public to follow new developments and participate in the public discourse on scientific issues (Ross, 1999).

A Social Justice Approach to Dealing with Environmental Crime

People often believe that responsibility for health rests entirely with the individual and therefore, public health threats such as AIDS, smoking, heart disease, and cancer are individual problems. Changing public opinion begins with informed education but awareness is just the beginning.

> "Effective public opinion is more than widespread awareness of a social problem, more than desire for change, more than a planned demonstration on a busy street corner designed to draw the attention of otherwise uninterested passersby. Instead, effective public opinion is that expression of sentiment that actually reaches the systematic agenda of political decision-makers" (Salmon and Christensen, 2003: 7).

A social justice approach means that researchers and activists work together to fight against environmental crime. It means examining the underlying structural causes of environmental crime. It means contributing to informed public participation efforts to eliminate environmental crime and injustice. There are a number of organizations and agencies which advocate a social justice approach to dealing with injustice. One such organization is the Citizen's Clearinghouse for Hazardous Waste.

Citizen's Clearinghouse for Hazardous Waste

The Citizen's Clearinghouse for Hazardous Waste (CCHW) is a non-profit organization founded in 1981 by Lois Gibbs, leader of the campaign at Love Canal. The CCHW is a national grassroots organization which strives to "translate scientific issues into plain language" (Gibbs, 1995: xxiii) and help activists fight for environmental and human health causes. In 1995, Gibbs published *Dying from Dioxin*. The first section of the book describes the health impacts of dioxin exposure with reference to EPA and several scientific studies. The second part of the book details how communities can organize and fight for their health. Gibbs emphasizes that just knowing the truth won't stop corporations from polluting the environment; community organization is necessary. She also points out that the EPA is not an ally; in reality the EPA protects the right to pollute by justifying standards that protect the interests of corporations and the government. According to Gibbs, successful organizations are community based, but nationally linked; they involve a large and diverse group of people; and offer up a clear and simple plan of action. The organization plan outlined in *Dying from Dioxin* (1995) is a must-read for grassroots activists. Gibbs knows that effective change begins at the community level:

> "We need to make it more expensive to pollute than it is to change. Corporations will not change behavior because CEOS wake up one morning and decide that stopping pollution is the right thing to do. Corporations will change because people-consumers, voters, and workers-convince them that they must change. This change will not come from Washington, D.C. It will not appear in the form of a top-down regulatory mandate. Changes in corporate behavior will only be accomplished through people working at the local level, then joining together at the state level, and then at the national level. Change depends on you, me, and millions of others who are willing to make that leap of faith from education into collective action" (Gibbs, 1995: 293).

The Time is Now

Every day, thousands of people die from various diseases and illnesses. Their deaths may be recorded as the result of a heart attack, cancer, stroke or the like. But there will be no parentheses beside the cause of death to indicate (lived near a toxic waste site) or (high toxic levels in drinking water). Their deaths will not be counted as homicides or negligent manslaughters. The media will not highlight their deaths in print or in newscasts aside from the obligatory obituaries. Every once in a while, if too many people living in close proximity to one another appear to be getting the same illnesses and same diseases, there might be some social, political, and media attention given to the community. If this community is disproportionately minority or low-income, the attention will be slow to come and action will be less likely to occur. Social, political, and media attention will only be maintained if the very people getting sick, who are also trying to raise families and support families, fight for the attention. No one else is going to come to the aid of those suffering unless they have something to gain from the assistance. If the local and federal governments are pushed hard enough to respond, then scientists will be called in to ascertain whether or not the illnesses and diseases are directly related to environmental contaminants or toxins. These scientific tests may take years and in the meanwhile, people living in these communities will continue to get sick. The majority cannot move away from the community. They do not have the financial resources to do so and their homes are losing value. The corporations responsible for the environmental toxins which are causing the community to suffer begin to cover their tracks and are able to spend millions to prepare defenses and to create PR campaigns that tout their environmentally friendly image. If tests eventually connect the human sickness and disease to the environmental toxins and the corporations responsible for their illegal (and sometimes legal) presence in the air, soil, or water, the battle is not over. In some cases, the corporation will be required to pay a fine. In most cases, the community will be forced to live with the consequences of the environmental contamination. The government rarely relocates communities due to environmental contamination. Rarely, if ever, will a corporate leader face criminal charges for his/her involvement. Social, political, and media attention will be short lived if given at all. And the people will continue to suffer and die.

We cannot continue to ignore environmental crime and its lethal consequences. There is abundant evidence that industrial pollutants cause a significant number of diseases and illnesses in the United States. Researchers from multiple disciplines need to share their findings and unite in the fight for our environment and health. Research isn't enough. We must also teach about environmental crime. We must give our time and expertise to the community. Community groups need the support of researchers to effectively challenge the power and financial influence of corporations. We must establish ties with local and national media to get our research into the public eye. We cannot waste anymore time. The time is now.

Appendix

Nation's Most Widely Circulated Newspapers 2004

Newspaper	Largest Reported Daily Circulation
The Atlanta Journal and Constitution (Georgia)	606,246
The Baltimore Sun (Maryland)	454,045
The Boston Globe (Massachusetts)	707,813
The Boston Herald (Massachusetts)	240,759
The Buffalo News (New York)	282,618
Chicago Sun-Times (Illinois)	963,927
The Columbus Dispatch	361,304
Daily News (New York)	786,952
The Denver Post/Rocky Mountain News (Colorado)	750,593
The Houston Chronicle (Texas)	737,580
Los Angeles Times (California)	1,292,274
Miami Herald (Florida)	416,530
The New York Times (New York)	1,680,583
Omaha World Herald (Nebraska)	242,964
Pittsburgh Post-Gazette (Pennsylvania)	402,981
San Diego Union-Tribune (California)	433,973
The San Francisco Chronicle (California)	540,314
The Seattle Times (Washington)	462,920
St. Louis Post-Dispatch (Missouri)	449,845
St. Petersburg Times (Florida)	395,973
Star Tribune (Minneapolis MN)	678,650
The Tampa Tribune (Florida)	293,090
The Times-Picayune (Louisiana)	281,374
USA Today (National)	2,665,815
The Washington Post (D.C.)	1,007,487

Source: Audit Bureau of Circulation
http://www.accessabc.com/reader/top100.htm

References

Adeola, F.O. 2000. Endangered community, enduring people: toxic contamination, health, and adaptive responses in a local context. *Environment and Behavior* 32: 209-249.

Albanese, J.S. and Pursley, R.P. 1993. *Crime in America: some existing and emerging areas.* Englewood Cliffs, NJ: Regents/Prentice Hall.

American Opinion Research, Inc. 1993. *The press and the environment: new journalists evaluate environmental reporting.* Princeton, NJ: Author.

American Petroleum Institute. 2004. *About oil and natural gas.* Washington, D.C.: American Petroleum Institute.

Anderson, A. and Gaber, I. 1993. The yellowing of the greens. *British Journalism Review* 4: 49-53.

Bachman, R. and Schutt, R.K. 2001. *The practice of research in criminology and criminal justice.* Thousand Oaks, CA: Pine Forge Press.

Bagdikian. 2000. *The media monopoly, sixth edition.* Boston, MA: Beacon Press.

Barak, G. 1988. Newsmaking criminology: reflections on the media, intellectuals, and crime. *Justice Quarterly* 5: 565-587.

Barak, G., ed. 1994. Media, process, and the social construction of crime: studies in newsmaking criminology. New York: Garland Pub.

Barlow, M.H. 1991. Ideologies of crime in the media: a content analysis of crime news in Time magazine in the Post-World War II period. Unpublished doctoral dissertation, School of Criminology and Criminal Justice, Florida State University.

Barlow, M.H., Barlow, D.E. and Chiricos, T.G. 1995. Economic conditions and ideologies of crime in the media: a content analysis of crime news. *Crime & Delinquency* 41: 3-19.

Barlow, M.H., Barlow, D.E. and Chiricos, T.G. 1995. Mobilizing support for social control in a declining economy: exploring ideologies of crime within crime news. *Crime & Delinquency* 41: 191-204.

Benedict, H. 1992. *Virgin or vamp: how the press covers sex crimes.* New York: Oxford University Press.

Berk, R.A., Brookman, H. and Lesser, S.L. 1977. *A measure of justice: an empirical study of changes in the California Penal Code 1955-1971.* New York: Academic Press.

Berkowitz, D. 1987. TV news sources and news channels: a study in agenda-building. *Journalism Quarterly* 64: 508-13.

Berkowitz, D. and Beach, D.W.1993. News sources and news context: the effect of routine news, conflict, and proximity. *Journalism Quarterly* 70: 4-12.

Berry, J. 1988. A long summer of smog. *Newsweek* 111: 46-48.

Bierne, P and Messerschmidt, J. 1991. *Criminology*. New York: Harcourt Brace Jovanovich.

Brown, J.K., Bybee, C.R., Wearden, S.T., and Straughan, D.M. 1987. Invisible power: newspaper news sources and the limits of diversity. *Journalism Quarterly* 64: 45-54.

Bryant, B. (ed.). 1995. *Environmental Justice: Issues, Policies, and Solutions*. California: Island Press.

Bullard, R. 1983. "Solid waste sites and the Houston black community." *Social Inquiry*, 53, 273-284.

Bureau of Justice Statistics. 1999. *Federal enforcement of environmental laws, 1997.* Washington, D.C.: U.S. Department of Justice, Office of Justice Programs.

Burns, R.G. and Lynch, M.J. 2004. *The sourcebook on environmental crime*. New York: LFB Publishers.

Calavita, K. and Pontell, H.N. 1994. The state and white collar crime: saving the savings and loans. *Law and Society Review* 28: 297-324.

Campbell-Mohn, C., Breen, B., and Futrell, J.W. 1993. *Sustainable environmental law*. St. Paul, MN: West.

Carson, R. 1962. *Silent Spring*. Boston, Houghton Mifflin Company.

Cavendar, G. 1984. *'Scared Straight': ideology and the media.* In Justice and the media: issues and research, edited by R. Surette. Springfield, IL: Charles C Thomas.

Chermak, S. 1994. Body count news: how crime is presented in the news media. *Justice Quarterly* 1: 561-82.

Chermak, S. 1995. *Victims in the news: crime and the American news media*. Boulder, CO: Westview.

Chermak, S. 1997. The presentation of drugs in the news media: the news sources involved in the construction of social problems. *Justice Quarterly* 14: 687-718.

Chermak, S. 1998. *Police, courts, and corrections in the news.* In F. Bailey and D. Hale, editors, Popular culture, crime and justice. Belmont, CA: Wadsworth.

Cheurprakobkit, S. and Bartsch, R.A. 2000. School crime and education: is there a need for criminology/criminal justice courses in the high school curriculum? *Journal of Security Administration* 23:1-12.

Children Now. 2003. Annual Report, 2003 highlights. Oakland, CA: Children Now.

Chiricos, T., Escholtz, S. and Gertz, M. 1997. Crime, news, and fear of crime: toward an identification of audience effects. *Social Problems* 44: 342-357.

Christensen, J., Schmidt, J., and Henderson, J. 1982. The selling of the police: media, ideology, and crime control. *Contemporary Crisis*: 6: 227-39.

Clifford, M. 1998. *Environmental crime: enforcement, policy, and social responsibility.* Gaithersburg, Maryland: Aspen Publishers, Inc.

Clinard, M.B. and Yeager, P.C. 1980. *Corporate Crime.* New York: Free Press.

Cockburn, A. 1997. Beat the devil. *Nation*, pages 9-10.

Cohen, B. 1963. *The press and foreign policy.* Princeton, NJ: Princeton University Press.

Cohen, S. 1975. A comparison of crime coverage in Detroit and Atlanta newspapers. *Journalism Quarterly* 52: 726-730.

Cohen, S. and Young, J., eds. 1981. *The manufacture of news.* Thousand Oaks, CA: Sage.

Cole, L.W. 1992. Empowerment as the key to environmental protection: the news for environmental poverty law. *Ecology Law Quarterly* 19: 619-683.

Combs, B. and Slovic, P. 1978. Newspaper coverage of causes of death. *Journalism Quarterly* 56: 837-43.

Compaine, B.M. and Gomery, D. 2000. *Who owns the media: competition and concentration in the mass media industry.* Mahwah, NJ: Lawrence Erlbaum Associates, Publishers.

Cope, G. 2002. Bush administration is lax on toxic clean-ups. *The Progressive* March 7th.

Council on Environmental Quality. 1995. *Environmental Quality: 1994-1995 report.* Washington, D.C.: Executive Office of the President

Currie, E. 1993. Reckoning: drugs, the cities, and the American future. New York: Hill and Wang.

Cushman, J.H. 1998. Pollution policy is unfair burden, states tell EPA. *New York Times* on the web. May 10.

Davis, D. 2002. *When smoke ran like water: tales of environmental deception and the battle against pollution.* New York, New York: Basic Books.

Day, D. 1989. *The environmental war.* New York: Ballantine.

Dearing, J.W. and Rogers, E.M. 1996. *Communication concepts 6: agenda-setting.* Thousand Oaks, CA: Sage.

DeLuca, K.M. 1999. *Image politics: the new rhetoric of environmental activism.* New York: The Guilford Press.

Department of Energy. 1997. Impact of environmental compliance costs on U.S. refining profitability. Washington, D.C.: Energy Information Administration.

Department of Energy. 1998. Overview and domestic and international trends for petroleum refining, 1998. Washington, D.C.: Energy Information Administration.

Department of Energy. 2003. Petroleum supply annual. Washington, D.C.: Energy Information Administration.

Department of Energy. 2003b. NPRA United States Refining and Storage Capacity Report. Washington, D.C.: Energy Information Administration.

Department of Energy. 2004. Energy information administration industry briefs. Washington, D.C.: Office of Industrial Technologies.

Department of Health, Education, and Welfare. 1980. *Report of the Subcommittee on the Potential Health Effects of Toxic Chemical Dumps on the DHEW Committee to Coordinate Environmental and Related Problems.* Washington, D.C.: Department of Health, Education, and Welfare.

Dowie, M. 1977. Pinto Madness. *Mother Jones*, September, 13-29.

Dowie, M. 1995. *Losing ground: American environmentalism at the close of the twentieth century.* Cambridge, MA: MIT Press.

Dussuyer, I. 1979. *Crime news: a study of 40 Ontario newspapers.* Toronto: University of Toronto Press.

EcoSystems. 2004. Hubbert Peak of Oil Production. Accessed on 9/4/2004 at http://www.hubbertpeak.com/summary.htm.

Edwards, S.M., Edwards, T.D., and Fields, C.B. 1996. *Environmental Crime and Criminality: Theoretical and Practical Issues.* New York: Garland Publishing, Inc.

Eisenbach, R. 2002. *The color of television: a multi-cultural look at the effects of television.* Eugene, OR: Media Literacy Online Project.

Entman, R.M. 1990. Modern racism and images of Blacks in local television news. *Critical Studies in Mass Communication* 7: 332-45.

Entman, R.M. 1992. Blacks in the news: television, modern racism, and cultural change. *Journalism Quarterly* 69: 341-61.

Entman, R.M. 1994. Representation and reality in the portrayal of Blacks on network television news. *Journalism Quarterly* 71: 509-20.

Environmental Protection Agency. 1995a. Profile of the petroleum industry. Office of Compliance Sector Notebook Project. Washington, D.C.: U.S. Government Printing Office.

Environmental Protection Agency. 1995b. Regulatory impact analysis for the petroleum refinery NESHAP. Research Triangle Park, North Carolina: Office of Air Quality Planning and Standards, Emission Standards Division.

Environmental Protection Agency. 2004a. Enforcement actions and tools. Washington D.C.: Office of Enforcement and Compliance Assurance.

Environmental Protection Agency. 2004b. Integrated compliance information system. Washington D.C.: Office of Enforcement and Compliance Assurance.

Environmental Protection Agency. 2004c. Petroleum refinery initiative. Washington, D.C.: Office of Compliance and Enforcement.

Environmental Protection Agency. 2004d. Enforcement case report: data dictionary. Washington, D.C.: Office of Compliance and Enforcement.

Envirotools. 2004. Petroleum factsheet. Accessed on 6/25/04 at www.envirotools.org/factsheets/petroleum.shtml

Ericson, R.V., Baranek, P.M., and Chan, J.B.L. 1987. *Visualizing deviance: a study of news organizations.* Toronto: University of Toronto Press.

Ericson, R.V., Baranek, P.M., and Chan, J.B.L. 1991. *Representing order: crime, law, and justice in the news media.* Toronto, Canada: University of Toronto Press.

Evans, S.S. and Lundman, R.J. 1983. Newspaper coverage of corporate pricefixing. *Criminology* 21: 529-41.

Federal Bureau of Investigation. 2003. *Environmental crime: facts and figures, 2003.* Accessed on May 17th, 2004 at http://www.fbi.gov

Fisher, G. 1989. Mass media effects on sex role attitudes of incarcerated men. *Sex Roles: A Journal of Research* 20: 191-203.

Fishman, M. 1978. Crime waves as ideology. *Social Problems* 25: 531-43.

Fishman, M. 1980. *Manufacturing news.* Austin, TX: University of Texas Press.

Foner, E. and Garraty, F. (eds.). 1991. The reader's companion to American history. Boston, MA: Houghton Mifflin Company.

Frank, N., and Lynch, M.J. 1992. *Corporate crime, corporate violence.* New York: Harrow and Heston.

Freudenrich, C.C. 2004. Oil refining. Accessed on 6/16/04 at www.howstuffworks.com/oil-refining.htm

Funkhouser, G.R. 1973a. The issues of the sixties: an exploratory study in the dynamics of public opinion. *Public Opinion Quarterly* 37: 62-75.

Funkhouser, G.R. 1973b. Trends in media coverage of the issues of the sixties. *Journalism Quarterly* 50: 533-538.

Gaber. 2000. *The greening of the public, politics and the press, 1985-1999.* In J. Smith (ed.), The daily globe: environmental change, the public and the media. London: Earthscan Publications Ltd.

Gans, H.J. 1979. *Deciding what's news.* New York: Vintage.

Gardiner, G.S. and McKinney, R.N. 1991. The great American war on drugs: another failure of tough-guy management. *Journal of Drug Issues* 21:605-16.

Garofalo, J. 1981. Crime and the mass media: a selective review of research. Journal of Research in Crime and Delinquency 18: 319-50.

Gaynor, M. 2002. In Davis, D. When smoke ran like water: tales of environmental deception and the battle against pollution. New York, New York: Basic Books.

Gerbner, G., Gross, L., Morgan, M. and Signorielli, N. 1994. *Growing up with television: the cultivation perspective.* In J. Bryant and D. Zillmann (eds). Media effects: advances in theory and research. Hillsdale, NJ: Lawrence Erlbaum Associates.

Gibbs, L.M. 1995. *Dying from dioxin.* Boston, Ma: South End Press.

Giddens, P.H. 1938. *The birth of the oil industry.* New York: The MacMillan Company.

Gitlin, T. 1980. *The whole world is watching: mass media and the making and unmaking of the new left.* Berkeley, CA: University of California Press.

Glaser, M. 1997. Censorious advertising. *Nation* page 7.

Goldman, B.A. 1991. *The truth about where you live.* New York: Times Books.

Goldstein, I.F. and Goldstein, M. 2002. *How much risk: a guide to understanding environmental health hazards.* New York: Oxford University Press.

Goodell, J. 1999. Death in the redwoods. *Rolling Stone,* pages 60-69.

Gordon, M. and Heath, L. 1981. *The news business, crime and fear.* In D. Lewis (ed.). Reactions to crime. Beverly Hills, CA: Sage.

Graber, D.A. 1980. *Crime news and the public.* New York: Praeger.

Grabosky, P. and Wilson, P. 1989. *Journalism and justice: how crime is reported.* Leichhardt, AUS: Pluto Press.

Graves, S.B. 1996. *Diversity on television.* In T.M. Macbeth (ed.), Tuning in to young viewers: social science perspectives on television. Thousand Oaks, CA: Sage.

Gray, C. 1998. Louisiana is nation's no. 2 polluter; No. 1 in amount of water waste. *The Times-Picayune,* page A2.

Greek, C. 1997. Using the Internet as a newsmaking criminology tool. Presented at the American Society of Criminology Annual Meetings, San Diego, CA.

Greer, J. and Bruno, K. 1996. *Greenwash: the reality behind corporate environmentalism.* New York: The Apex Press.

Hall, S., Critcher, C., Jefferson, T., Clarke, J., and Roberts, B. 1978. *Policing the crisis: mugging, the state, and law and order.* London: Macmillan.

Hansen, A. 1993. *The mass media and environmental issues.* Leicester University Press.

Haskins, J.B. and Miller, M.M. 1984. The effects of bad news and good news on a newspaper's image. *Journalism Quarterly* 61: 3-13.

Hearold, S. 1986. *A synthesis of 1043 effects of television on social behavior.* In G. Comstock (ed.). Public communication and behavior. Orlando, FL: Academic Press.

Heath, L. 1984. Impact of newspaper crime reports on fear of crime: multimethdological investigation. *Journal of Personality and Social Psychology* 47: 263-276.

Heath, L. and Gilbert, K. 1996. Mass media and fear of crime. *American Behavioral Scientist* 39: 379-386.

Helvarg, D. 1994. *The war against the Greens: the wise-use movement, the new right, and anti-environmental violence.* San Francisco, CA: Sierra Club Books.

Herman, E.S. and Chomsky, N. 1988. *Manufacturing consent: the political economy of the mass media.* New York: Pantheon Books.

Hertsgaard, M. 1990. Covering the world: ignoring the earth. *Greenpeace,* pages 14-18.

Hills, S.L. 1987. *Corporate violence: injury and profit for death.* Totowa, NJ: Rowman and Littlefield.

Humphries, D. 1981. Serious crime, news coverage, and ideology: a content analysis of crime coverage in a metropolitan newspaper. *Crime and Delinquency* 27: 191-205.

Humphries, S.L. 1990. An enemy of the people: prosecuting the corporate polluter as a common law criminal. *American University Law Review* 39: 311-354.

Jacobson, L. 1998. Wilderness crusade. *Washington City Paper,* page 48.

Jensen, E.L., Gerber, J. and Babcock, G.M. 1991. The new war on drugs: grassroots movement or political construction? *Journal of Drug Issues* 21: 651-67.

Jerin, R.A. and Fields, C.B. 1995. *Murder and mayhem in USA Today: a quantitative analysis of the national reporting of states' news.* In G. Barak (ed.), Media, process, and the social construction of crime: studies in newsmaking criminology. New York: Garland.

Johnstone, J. W.C., Hawkins, D.F., and Michener, A. 1994. Homicide reporting in Chicago dailies. *Journalism Quarterly* 71: 860-72.

Kappeler, V., Blumberg, M., and Potter, G.W. 1996. *The mythology of crime and criminal justice, second edition.* Prospect Heights, IL: Waveland.

Kessel, J.H. 1975. *The domestic presidency: decision-making in the White House.* North Scituate, MA: Duxbury.

Kuehn, R. 1996. The environmental justice implications of quantitative risk assessment. University of Illinois Law Review.

Laswell, H. 1948. *The structure and function of communication in society.* In L. Bryson (ed.). The communication of ideas. New York, Harper.

Lavelle, M. 1993. Environmental vise: law, compliance. *National Law Journal,* August 30: S8.

Lavelle, M., and Coyle, M. 1992. "Unequal protection: The racial divide in environmental law." In Richard Hofrichter (ed.), Toxic struggles: The theory and practice of environmental justice (pp. 136-143). Philadelphia, PA: New Society Publishers.

Lazarsfeld, P.F. and Merton, R.K. 1948. *Mass communication, popular taste, and organized social action.* In L. Bryson (ed.). The communication of ideas. New York, Harper.

Leonard, T. 1986. The power of the press: the birth of American political reporting. New York: Oxford University Press.

Levi, M. 1994. The media and white collar crime. Presented at the annual meetings of the American Society of Criminology, Miami.

Lichenstein, S., Slovic, P., Fishoff, B., Layman, M., and Combs, B. 1978. Judged frequency of lethal events. *Journal of Experimental Psychology*: Human Learning and Memory 4: 551-78.

Lictblau, J.H. 1992. Issues affecting the refining sector of the petroleum industry. New York: Petroleum Industry Research Foundation, Inc.

Linsky, M. 1988. *The media and public deliberation.* In The Power of public ideas, edited by R. Reich. Cambridge, MA: Harvard University Press.

Lippmann, W. 1922. *Public opinion.* New York: Macmillan.

Liska, A.E. and Baccaglini, W. 1990. Feeling safe by comparison: crime in the newspapers. *Social Problems* 37: 360-374.

Lofquist, W. 1997. Constructing 'crime': media coverage of individual and organizational wrongdoing. *Justice Quarterly* 14: 243-263.

Lotz, R. 1991. *Crime and the American press.* New York: Praeger.

Lynch, M.J. 1990. The greening of criminology: A perspective on the 1990s. *Criminologist,* 2: 11-12.

Lynch, M.J. and Patterson, E.B. 1991. *The biases of bail: race and gender discrimination in formalized bail procedures.* Paper presented at the annual meetings of the Academy of Criminal Justice Sciences, Nashville, TN, March 1991.

Lynch, M.J. and Patterson, E.B., eds. 1996. *Justice with prejudice.* Albany, NY: Harrow and Heston.

Lynch, M.J., McGurrin, D. and Fenwick, M. 2004. Disappearing act: the representation of corporate crime research in the criminological literature. *Journal of Criminal Justice* 32: 389-398.

Lynch, M.J., Nalla, M.K., and Miller, K.W. 1989. Cross-cultural perceptions of deviance: the case of Bhopal. *Journal of Research in Crime and Delinquency* 26: 7-35.

Lynch, M.J., Stretesky, P.B. and Hammond, P. 2000. Media coverage of chemical crimes, Hillsborough County, Florida, 1987-97. *The British Journal of Criminology* 40:112-26.

Lynch, M.J., P. Stretesky and D. McGurrin . 2001. *Toxic Crimes and Environmental Justice.* In G. Potter's (ed), Controversies in White Collar Crime. Cincinnati : Anderson.

Lynch, M.J. and Stretesky, P. 2001. Toxic crimes: examining corporate victimization of the general public employing medical and epidemiological evidence. *Critical Criminology* 10: 153-172.

Lynch, M.J., Stretesky, P.B. and Burns, R.G. 2004a. Slippery business: race, class, and legal determinants of penalties against petroleum refineries. *Journal of Black Studies* 34: 421-440.

Lynch, M.J., Stretesky, P.B. and Burns, R.G. 2004b. Determinants of environmental law violation fines against oil refineries: race, ethnicity, income, and aggregation effects. *Society and Natural Resources* 17: 333-347.

MacDonald, Z. 2002. Official crime statistics: their use and interpretation. *The Economic Journal* 112: 85-106.

MacKuen, M.B. and Coombs, S.L. 1981. *More than news: media power in public affairs.* Beverly Hills, CA: Sage.

Maguire, Brendan, "Television Network News Coverage of Corporate Crime from 1970-2000." Western Criminology Review 3 Online, Available: http://wcr.sonoma.edu/v3n2/maguire.html.

Manoff, R. and Schudson, M. 1986. *Reading the news.* New York: Pantheon Books.

Marsh, H.L. 1989. Newspaper crime coverage in the U.S.: 1893-1988. *Journal of Criminal Justice* 19: 67-78.

Mayo, D.G. and Hollander, R.D. (eds.). 1991. *Acceptable evidence: science and values in risk management.* New York: Oxford University Press.

McCombs, M.E. and Shaw, D.L. 1972. The agenda-setting function of the mass media. *Public Opinion Quarterly* 36: 176-187.

McDonald, S. 1993. The media and misinformation: how the press has fueled an anti-environmental backlash. *Friends of the Earth*, pages 6-7.

Mediamark Research Inc. 2003. *Newspaper section readership 2004.* NAA Business Analysis and Research Department.

Michalowski, R.J. 1985. *Order, law, and crime.* New York: Random House.

Miller, M.C. 1996. Free the media. *Nation*, pages 9-15.

Miller, M.C. 1998. TV: the nature of the beast. *Nation,* pages 11-13.

Mohai, P., and Bryant, B. 1992. *Environmental racism: Reviewing the evidence.* In Bunyan Bryant and Paul Mohai (eds.), Race and the incidence of environmental hazards: A time for discourse (pp. 163-175). Boulder, CO: Westview Press.

Morash, M. and Hale, D. 1987. *Unusual crime or crime as usual? Images of corruption at the Interstate Commerce Commission.* In

Organized Crime in America: Concepts and Controversies, edited by T.S. Bynum. Monsey, NY: Criminal Justice Press.

Nadakavukaren, A. 1995. *Our global environment: a health perspective.* Prospect Heights, IL: Waveland Press.

National Academy of Sciences. 2003. Cumulative environmental effects of oil and gas activities on Alaska's North Slope. Washington, D.C.: National Research Council.

National Association of Broadcasters. 1995. Radio activities. Available at http://www.nab.org/www/userguid/radio.htm

National Resources Defense Council. 2001. Oil companies, making record profits, seek environmental rollbacks. Washington, D.C.: Environment Media Services.

National White Collar Crime Center. 2004. Environmental Crime Defined. Accessed May 2004 at http://www.nw3c.org/index.html.

Nelkin, D. and Brown, M.S. 1984. *Workers at risk: voices from the workplace.* Chicago: University of Chicago Press.

Neuzil and Kovarik. 1996. *Mass media and environmental conflict.* Thousand Oaks, CA: Sage.

Newman, G. 1990. Popular culture and criminal justice: a preliminary analysis. *Journal of Criminal Justice* 18: 261-74.

Noelle-Neumann, E. 1991. *The theory of public opinion: the concept of the spiral of silence.* In J.A. Anderson (ed.). Communication yearbook, volume 14. Newbury Park, CA: Sage.

Noelle-Nuemann, E. 1993. *The spiral of silence: public opinion- our social skin.* Chicago, IL: University of Chicago Press.

Nisbett, R. and Ross, L. 1980. *Human inference: strategies and shortcomings of social judgment.* Englewood Cliffs, NJ: Prentice-Hall.

Occupational Safety and Health Administration. 2004. Petroleum refining. Washington, D.C.: U.S. Department of Labor.

Owen, J.W. 1975. Trek of the oil finders. Tulsa, OK: American Association of Petroleum Geologists.

Owen, O.S. 1975. *Natural resource conservation, second edition.* New York: MacMillan.

Paik, H. and Comstock, G. 1994. The effects of television violence in antisocial behavior. *Communication Research* 21: 516-546.

Parenti, *M. 1993.* Inventing reality. *New York: St. Martin's Press.*

Parenti, M. 1997. *The hidden ideology of the mass media.* Seattle, WA: Peoples Video/Audio.

Pedhazur, E. J. 1997. *Multiple Regression in Behavioral Research (3rd ed.).* Orlando, FL: Harcourt Brace.

Pees, S.T. 2004. Oil history. Accessed on 6/25/04 at
www.oilhistory.com

Perse, E.M. 2001. *Media effects and society.* Mahwah, NJ: Lawrence
Erlbaum Associates, Publishers.

Pew Research Center for the People and the Press. 2000. *Self-
Censorship: how often and why- journalists avoiding the news.*
Accessed on March 11, 2005 at www.people-press.org.

Ponder, S. 1986. Federal news management in the progressive era:
Gifford Pinochot and the conservation crusade. *Journalism
History* 13: 42-48.

Pope, C. 2004. Strategic ignorance: why the Bush Administration is
recklessly destroying a century of environmental progress. Sierra
Club Books.

Potter, C.B. 1998. War on crime: bandits, G-men, and the politics of
mass culture. New Brunswick, NJ: Rutgers University Press.

Pritchard, D. 1986. Homicide and bargained justice: the agenda-setting
effect of crime news on prosecutors. *Public Opinion Quarterly* 50:
143-59.

Randall, D. M. 1987. The portrayal of business malfeasance in the elite
and general public media. *Social Science Quarterly* 68: 281-93.

Randall, D. and Defillippi, R. 1987. Media coverage of corporate
malfeasance in the oil industry. *The Social Science Journal* 24: 31-
42.

Randall, D. and Lee-Sammons, L. 1988. Common versus elite crime
coverage in network news. *Social Science Quarterly* 69: 910-29.

Reiman, J. 1998. *The rich get richer and the poor get prison.* Boston:
Allyn and Bacon.

Reinarman, C. and Levine, H.G. 1995. *The crack attack: America's
latest drug scare, 1986-1992.* In J. Best (ed.), Images of issues:
typifying contemporary social problems. New York: Aldine.

Reske, H.J. 1992. Record EPA prosecutions. *ABA Journal*, March
1992, 25.

Roberts, T.J. and Toffolon-Weiss, M. 2001. *Chronicles from the
environmental justice frontline.* New York: Cambridge University
Press.

Rodman, G. 2001. *Making sense of the media.* Needham Heights, MA:
Allyn and Bacon.

Ross, S.S. 1999. *Coverage of women's health issues by the mass
media.* A study for the National Council on Aging, January 1998-
March 1998.

Ruggiero, G. and Sahulka, S., eds. 1999. *Project censored: the progressive guide to alternative media and activism.* New York: Seven Stories Press.

Sacco, V.F. 1995. Media constructions of crime. *The Annals of the American Academy of Political and Social Science* 539:141-54.

Sale, K. 1993. *The green revolution: the American environmental movement, 1962-1992.* New York: Hill and Wang.

Salmon, C.T. and Christensen, R.E. 2003. *Mobilizing public will for social change.* Lansing, MI: Michigan State University.

Schiller, D. 1981. *Objectivity and the news: the public and the rise of commercial journalism.* Philadelphia, PA: University of Pennsylvania Press.

Schlesinger, P., Tumber, H., and Murdock, G. 1991. The media politics of crime and criminal justice. *The British Journal of Sociology* 42: 397-420

Schudson, M. 1978. *Discovering the news.* New York: Basic Books.

Sheley, J.F. and Ashkins, C.D. 1981. Crime, crime news, and crime views. *Public Opinion Quarterly* 45: 492-506.

Sherizen, S. 1978. *Social creation of crime news: all the news fitted to print.* In Deviance and Mass Media, edited by C. Winick. Beverly Hills, CA: Sage.

Sherman, L. 1998. Needed: better ways to count crooks. *Wall Street Journal*, December 3rd.

Sigal, L.V. 1973. *Reporters and officials.* Lexington, MA: Heath.

Signorielli, N. 1990. *Television and health: images and impact.* In C. Atkin and L. Wallack (eds.), Mass communication and public health: complexities and conflicts. Newbury Park, CA: Sage.

Signorielli, N. 1993. *Television, the portrayal of women, and children's attitudes.* In G.L. Berry and J.K. Asamen (eds.), Children and television: images in a changing sociocultural world. Newbury Park, CA: Sage.

Simon, D.R. and Eitzen, D.S. 1993. *Elite deviance.* Boston, MA: Allyn and Bacon.

Simon, D.R. 2000. Corporate environmental crimes and social inequality: new directions for environmental justice research. *The American Behavioral Scientist* 43: 633-45.

Sissell, K. 1999. Equity programs strain state resources. *Chemical Week* July 28: 6.

Situ, Y. and Emmons, D. 2000. *Environmental crime: the Criminal Justice System's role in protecting the environment.* Thousand Oaks, CA: Sage.

Sjogren, H. and Skogh, G. 2004. *New perspectives on economic crime.* Cheltenham, U.K.: Edward Elgar Publishing, Inc.

Skogan, W.G. and Maxfield, M. 1981. *Coping with crime: victimization, fear, and reactions to crime.* Beverly Hills, CA: Sage.

Smith, S.J. 1984. Crime in the news. *British Journal of Criminology* 24: 289-295.

Stauber, J.C. and Rampton, S. 1995. Toxic sludge is good for you! Lies, damn lies, and the public relations industry. Common Courage Press.

Stempel, G.H. and Hargrove, T. 1996. Mass media audiences in a changing media environment. *Journalism and Mass Communication Quarterly* 73: 549-558.

Stretesky, P. and Lynch, M.J. 1999. Corporate Environmental Violence and Racism. *Crime, Law and Social Change* 30: 163-184.

Strodthoff, G.G., Hawkins, R.P. and Schoenfeld, A.C. 1985. Media roles in a social movement: a model of ideology diffusion. *Journal of Communication* 35: 135-153.

Stroman, C. and Seltzer, R. 1985. Media use and perceptions of crime. *Journalism Quarterly* 62: 340-45.

Surette, R. 1992. *Media, crime, and criminal justice : images and realities.* Pacific Grove, CA: Brooks/Cole Publishing Company.

Surette, R. 1998. *Media, crime, and criminal justice: images and realities, second edition.* Belmont, CA: West/Wadsworth.

Swigert, V.L. and Farrell, R.A. 1980. Corporate homicide: definitional processes in the creation of deviance. *Law and Society Review* 15: 161-82.

Taylor, D.E. 2000. The rise of the environmental justice paradigm: injustice framing and the social construction of environmental discourses. *American Behavioral Scientist* 43: 508-80.

Tonry, M. 1995. *Malign neglect: race, crime, and punishment in America.* New York: Oxford University Press.

Tunnell, K. 1992. Film at eleven: recent developments in the commodifications of crime. *Sociological Spectrum* 12: 293-313.

United Church of Christ. 1987. Toxic Wastes and Race: A National Report on the Racial and Socio-Economic Characteristics of Communities with Hazardous Waste Sites. New York: United Church of Christ.

U.S. Census Bureau. 1997. Petroleum refineries 1997 economic census. Washington, D.C.: U.S. Department of Commerce, Economics and Statistics Administration.

U.S. Census Bureau. 2001. Home computers and Internet use in the United States, 2000. Washington, D.C.: U.S. Department of Commerce, Economics and Statistics Administration.

U.S. Census Bureau. 2001. Annual survey of manufacturers, 2001. Washington, D.C.: U.S. Department of Commerce.

U.S. Chamber of Commerce. 1998. VA-HUD Appropriations Bill. Press Release October 5.

U.S. General Accounting Office. 1983. Siting of Hazardous Waste Landfills and Their Correlations with Racial and Economic Status of Surrounding Communities. Washington, D.C.: U.S. Government Printing Office.

U.S. Public Interest Research Group. 1999. Dirty dollars, dirty air: the auto and oil industries' continuing campaign against air pollution control. Washington, D.C.: PIRG.

Whelan, E. 1985. *Toxic terror: the truth about the cancer scare.* Ottawa, IL: Jameson.

White, R. 2003. Environmental issues and the criminological imagination. *Theoretical Criminology* 7: 483-506.

Wilbanks, W. 1984. *Murder in Miami: an analysis of homicide patterns and trends in Dade County, Florida, 1917-1983.* New York: University Press of America.

Williams, P. and Dickinson, J. 1993. Fear of crime: read all about it? The relationship between newspaper crime reporting and fear of crime. *Journal of Criminology* 33: 33-56.

Wilson, J.D. 1986. Rethinking penalties for environmental offenders: a view of the law reform commission of Canada's sentencing in environmental cases. *McGill Law Journal* 31: 315-332.

Wood, W., Wong, F.Y., and Chachere, J.G. 1991. Effects of media violence on viewers' aggression in unconstrained social interaction. *Psychological Bulletin* 109: 371-383.

Wright, C.R. 1986. *Mass communication: a sociological perspective.* New York: Random House.

Wright, J.P., Cullen, F.T., and Blankenship, M.B. 1995. The social construction of corporate violence: media coverage of the Imperial Food Products fire. *Crime & Delinquency* 41: 20-3

Index